Vortex/t

Charles D. Minahen

Vortex/t

THE POETICS OF TURBULENCE

The Pennsylvania State University Press

University Park, Pennsylvania

Library of Congress Cataloging-in-Publication Data

Minahen, Charles D.
Vortex/t : the poetics of turbulence / Charles D. Minahen.

p. cm.
Includes bibliographical references and indexes.
ISBN 0-271-00774-5
1. Vortex-motion in literature. 2. Turbulence in literature.
3. Symbolism in literature. 4. Literature—History and criticism.
I. Title. II. Title: Vortext. III. Title: Vortex.
PN56.V57M56 1992
809'.915—dc20 91-2243
CIP

Printed in the United States of America

It is the policy of The Pennsylvania State University Press to use acid-free paper for the first printing of all clothbound books. Publications on uncoated stock satisfy the minimum requirements of American National Standard for Information Sciences—Permanence of Paper for Printed Library Materials, ANSI Z39.48-1984.

Contents

PREFACE ix

PART I: Pattern in the Flux
1 Archaeology of Symbolic Turbulence 3
2 Cosmic Vortices of the Pre-Socratic Philosophers 15
3 Plato's "Great Whorl" 27
4 Epicurean Whirlings and Lucretius's Turbulent Flux 37

PART II: Visionary Breakthrough
5 Dante's Vortical Triptych 51
6 The Turbulent Dream-Vision of Descartes's "Olympian" Experience 71
7 ". . . That Every Thing Has Its / Own Vortex . . ." Dialectics of Vortical Symbolism in Blake 85

PART III: Threshold of the Unknown
8 Descents into Poe's Maelstrom 101
9 "Tourbillons de lumière": Rimbaud's Illuminating Vortices 115
10 Whirling Toward the Void at Dead Center: Symbolic Turbulence in Mallarmé's *Un Coup de dés* 129

CONCLUSION 143

APPENDIX: Vortices, Helices, Spirals, and Gyres 149

NOTES 171
WORKS CITED 193
INDEX OF SIGLA 199
INDEX 201

For
Marjory McCamley Minahen

Preface

Every great imaginative conception is a vortex into which everything under the sun may be swept.

—*J. L. Lowes,* The Road to Xanadu

Tout objet naissant est, d'abord, tourbillon, ainsi que le monde, en somme.

—*Michel Serres,* La Naissance de la physique

Turbulence—whirling vortices arising and dissipating unpredictably in a chaotic flux of randomly colliding currents—has for centuries attracted the interest of scientists intrigued by the paradox of its shifting states of order and disorder that seem all but impossible to describe, determine, and predict. Less apparent perhaps has been a very ancient tradition of aesthetic turbulence, involving symbolic representations of whirling spiro-helical and vortical phenomena, that originates in prehistory and develops across the ages into modern times. Even the symbols of the earliest humans reveal a fascination with turbulence and the recognition in it of profound, universal paradigms of nature and human experience that charge these images with cosmic, mystical significance.

This book aims at a hermeneutical disclosure of symbolic turbulence. By analyzing, comparing, and contrasting salient examples of vortical symbolism, I hope to "dis-cover" the dynamics of emerging and receding meanings that constitute the paradigm's polyvalence and complexity. The approach is diachronic to the degree that manifestations of the symbol are addressed chronologically, but my intention is not to suggest a historical concatenation of cause and effect, even if some such links do appear. I am more interested in reconstructing a synchrony of the symbol that places each discrete example of it in a vibrant intertext of patent and latent significations. It is obvious from these remarks that some of the book's fundamental theoretical and organizational strategies have been derived from phenomenology. My subject involves symbolic representations and transformations of a natural phenomenon and thus lends itself readily to a phenomenological approach. But I have felt in no way bound to any set corpus of concepts or codes and, depending on the specifics of each particular case, have referred to and applied a variety of critical theories and techniques.

The idea for this study grew out of my research into various *symboles privilégiés* in the poetry of Rimbaud and Mallarmé, who are the subjects, respectively, of Chapters 9 and 10. As I looked for antecedents to symbols of turbulence in their works, I found myself working my way back to the very origins of symbolism, constantly surprised and delighted by the authors and works that offered new discoveries. Given my focus on symbols, I sought out primarily poetic expressions of turbulence, not expecting to find them in authors like Heraclitus, Plato, and Descartes. But the symbols that they set forth are unmistakably aesthetic in their appeal to pure imagination, and when so expressing themselves, these great thinkers are in every way poets. Most of the other examples analyzed are more obviously poetic, although it is not always a question of verse, since it is the privileged use of the symbol—the most essential characteristic of poetry, in my view—that has dictated the choice of texts, admitting even illustrations in the unique case of Blake.

In order to provide the reader with an introduction to the background and the vocabulary of turbulence, including explanations of concepts and terms, descriptions of structures, and concrete examples of these when possible, I have appended a chapter that attempts to bring clarity and consistency to a field that, when I began my research, seemed utterly confused and contradictory. The rest of the book assumes many of the definitions and explanations provided there, so at least an initial glance through those pages is advised.

Since many of the primary and secondary texts I have analyzed are in French, I have translated them into English, in order to make them accessible to non-French speakers. In most cases English has replaced the French, but when the French text is itself the object of analysis—the case primarily in Chapters 6, 9, and 10—I have placed the English translation in brackets following the French. I have used brackets instead of parentheses in these instances to simplify the reading and to avoid confusion with parenthetical elements that are not translations. Unless otherwise noted, the translations are my own and reflect a preference for the literal meaning over what may have sounded better in English. They are thus context specific and are not meant to stand alone as polished finished products.

In the matter of isolated cognates, I have tended not to translate them if their meaning is obvious, and I have omitted reproducing in the translations italics, unusual spacing or layout, and other special effects that are apparent in the original. For texts in languages other than French, I have tried to select the best English translations, often availing myself of more than one. In some instances, the untranslated vocabulary of turbulence has been interpolated to give the reader at least some direct access to the original.

An index of sigla has been assembled for quick identification of references abbreviated in parentheses. As a general rule, the first letter of each siglum is that of the author's last name and the second a letter taken from one of the principal words of the title. Full citations of sigla are given in Works Cited.

The chapters on Rimbaud, Mallarmé, and Descartes are revised and updated versions of articles previously published in, respectively, *Stanford French Review* 9:3 (Winter 1985): 351–64, published by the Department of French and Italian, Stanford University and Anma Libri; *The Romanic Review* 78:1 (January 1987), Copyright by the Trustees of Columbia University in the City of New York; and *Symposium* 41:2 (Summer 1987), reprinted with permission of the Helen Dwight Reid Educational Foundation (published by Heldref Publications, 4000 Albemarle St., N.W., Washington, D.C. 20016. Copyright © 1987). I thank these journals for permission to include material from the articles in this book. The College of Humanities and the Office of Research and Graduate Studies of the Ohio State University have provided a generous publication award for which I am grateful. I would also like to acknowledge the invaluable contribution of Karl Zamboni, who has been an enthusiastic supporter of this project since its inception and whose wealth of knowledge and shrewd bibliographical skills have proven enormously helpful in tracking down the many leads that have sprung from the delightful conversations we have had. I am similarly indebted to Raymond Giraud for his wise counsel and inspiring open-mindedness during particularly critical stages as the work took shape. Ella Kirk and Margaret Bolovan have provided excellent research and word-processing assistance, and I thank them for it. I have also greatly benefited from the encouragement of Patricia Servatius, whose faith in my ability to complete this task helped me persevere. Finally, I have very much appreciated David Prout's expert editorial advice and the infinite patience of Philip Winsor, Senior Editor at Penn State Press, whose support at a crucial moment really made a difference.

Part I

Pattern in the Flux

Archaeology of Symbolic Turbulence

1

What are the origins of symbolic turbulence? At least two very ancient symbolic types display, as an essential property, the dynamic whirl that is common to all the symbols of turbulence we shall examine. The first of these is a curvilinear whorl known as the "Tao symbol" or "yin-yang symbol of China,"[1] consisting of a circle divided by a sigmoid arabesque into opposing comma-shaped halves (magatamas), one dark (yin), the other light (yang), with a point of the opposite shade as a central "eye" in the "head" of each magatama. The second type, exemplified by various forms of the gammadion and triskelion, is essentially rectilinear and composed of straight line segments radiating from a center, like spokes of a wheel, with trailing appendages. Both types had mystical cosmic and religious significance, and both are structurally dialectical in their embodiment of polar opposition and synthesis.

In the case of the second type, the dialectic involves a dynamic fusion of rectilinearity and circularity, accomplished by the implied whirling of the radii about the figure's fixed center. Obviously, it is related to such symbols as the square inscribed within a circle or the circle inscribed within a square, which express the human desire to reconcile such limiting earthly realities as mortality, imperfection, and change (symbolized by the linear, segmented angularity of the square) with the eternal, immutable perfection of the divine circle, which has neither beginning nor end.[2] In this sense, the turbulent whirl effects a mystical symbolic convergence of earthly rectilinearity and heavenly curvilinearity.

The dialectic of the yin-yang whorl is similarly dynamic and also involves a spin, but the opposition is represented in terms of dark and light. Like its rectilinear counterpart, it embodies a principle of transformation between contraries, although here the opposites of the antinomies are already nascent within them. According to Stiskin, each magatama is a spiral (117) tracing a whirling path from the circle's circumference to the figure's central eye (where metamorphosis into the opposite is beginning to occur) or vice versa from the center (where the change presumably has already taken place) to the periphery. The Tao symbol thus uses the convolutions of a spiralic whirl to synthesize the opposition of center and periphery and to effect the transformation of light into dark and dark into light.

From this brief analysis of two ancient symbolic types, it is possible to derive the concepts of polar opposition, dynamic interaction, synthesis, and mystical transformation that emerge as essential properties of many of the symbols of turbulence selected for scrutiny in this study. Although the focus will be increasingly upon the vortex symbol, it is already clear that symbolic turbulence involves—and any analysis of it must take into consideration—certain tangentially related phenomena, such as the point, the line, the curve, the center, the circle, and to a very significant degree, the spiral, a fundamental form that requires further elaboration.

The spiral, which is a structural property of the Tao symbol, as we have just seen, is an even more universal and pervasive symbol in its own right and also seems to have possessed from ancient times occult cosmic and metaphysical significance.[3] It is found at various stages of human development (beginning with the earliest) among widely dispersed cultures, including ancient Egypt, Greece, Crete, and Troy, then later in Central Europe, Scandinavia, the British Isles, the Near and Far East, Indonesia and Polynesia, and pre-Columbian America. Mackenzie relates the debate between the "evolutionists" and "diffusionists" concerning the origin of the symbol as, respectively, independent and spontaneous or migrating from a single source.[4] Like so many matters connected with early humans, there is insufficient evidence to conclude one way or the other.

Whether the spiral was merely a decorative art motif devoid of any deeper significance or an authentic symbol charged with symbolic multiplesense that varied from culture to culture is another debated issue. Mackenzie himself argues vigorously for the symbolist position and succeeds in undermining the claims of the pure "ornamentalists." The problem here is that, unlike the cross and the gammadion, which originally were abstract "expressions of ideas regarding natural phenomena, natural processes," the spiral "was very certainly imitative" (MM 66), since it abounds in nature and is a pattern likely to have been admired and used ornamentally due to its aesthetic appeal. While this may be true, it does not preclude its use as a symbol, for it seems evident that it was generally considered "lucky" (MM 48), if not sacred, as appears to be the case when directly associated with an object or placed in a context now known to have had religious significance. When traced in bold relief against a contrasting background, the spiral, as a result of a riveting optical illusion, appears literally to gyrate, with radial lines spinning vibrantly about the center, an effect other rotational emblems do not display. Such a unique "magical" property obviously invests it with compelling mystico-religious appeal.

A case in point that illustrates the ambivalence concerning the decorative and/or symbolic function of the spiral is the rolled-up scroll or "volute" employed by the Greeks in the Ionic capital (where a complementary pair constitutes the principal motif) and the Corinthian capital (with its eight smaller volutes under the abacus). At first glance, the form's decorative role may seem paramount, since it is undeniable that the volutes "adorn" the columns in both instances, but the columns themselves are typically found supporting temples and other sacred edifices, so the motif may also have had religious symbolic significance, even if its meaning has not always survived.

In fact, as will become increasingly evident in the ensuing pages, the very structure of the spiral conduces to the representation of certain fundamental physical and metaphysical principles and beliefs. We have already seen, in the case of the Tao symbol, that the spiral traces a convoluted path that links, through a series of cycles, the center and the periphery (beginning and end). The movement toward or away from the center is thus progressive, making it a suitable symbol of "change" that "proceeds by degrees" (SL 25), and the vector of motion is reversible, since it may be centripetal or centrifugal, which necessarily implies a polar oppositeness of clockwise (right-handed) and counterclockwise (left-handed) gyration. Finally, the frequency of revolution along a spiral path decreases in proportion to movement away from the center, an effect that is realized gradually in the case of the tightly coiled Archimedean type (which extends itself cautiously and conservatively into space), while more abruptly and aggressively in the case of the equiangular form (which leaps boldly, at a logarithmic rate of distance between coils, to occupy more space more rapidly).

I shall not attempt here to explore in depth these characteristics, since the specific symbols we shall encounter provide ample opportunities for analysis. My purpose at this early stage is to establish the relevance and importance of the spiral to a study of symbolic turbulence and to draw attention to its rich symbolic potential. The same is all the more true for the three-dimensional spiral helix, which extends properties of the two-dimensional spiral and which is even more closely related structurally to the vortex symbol.

When the spiral's convolutions are translated along an axis, the whorled conical configuration of the spiral helix is produced. Many of the most common embodiments of this form are marine phenomena such as the seashell, the "magico-religious" use of which Mackenzie has discovered in "Upper Palaeolithic" cultures of Western Europe. With regard to the cryptic archeology of spiral symbolism, he concludes, "it may be that the shell spiral came first (66)."[5] Certainly its sculptural, brilliantly colored, and lustrous beauty lends it characteristics of a "jewel born of the sea,"[6] and its use as jewelry to adorn the human body testifies to its decorative value from ancient through modern times.

But a rich symbolism also inheres in the shell, which is linked in many ways to procreation, birth, and growth. Both English "shell" and French *coquille* denote a vessel that harbors burgeoning life (true not only of the seashell but of the eggshell and the nutshell, too). In this sense it is a matrix, like the female womb, and Eliade notes the "resemblance of the seashell to the female genital organs" (164), a purely symbolic correspondence that, far from being limited to a particular cultural group, is universally recognizable. Thus, the spiro-helical nature of birth symbolism, further reflected in the helical structure of the umbilical cord and "the necessary spiral movement before birth" (MM 111), is complemented by, says Eliade, "sexual symbolism" (164). The fleshy tone and texture of certain seashells, as well as the viscosity associated with both the sexual act and the soft " 'mollusk [which] exudes its shell,' lets the building material 'ooze out,' 'distills rhythmically its marvelous covering,' "[7] render the seashell, additionally, "a symbol of fertility" (EI 175). The supposed aphrodisiac properties of both oysters and pearls are relevant here, as is the myth of Aphrodite herself, "born from a marine conch" (EI 172).[8]

Bachelard, who recognizes in the *coquille* "spiral helices" (105n),[9] attempts to elicit "a shell dialectic" by interfacing its contradictory properties. First, there is the immediately evident opposition of "hard" and "soft" between the crusty shell and the tender creature it contains, which makes one wonder "how the softest being constitutes the hardest shell" (115). A second dialectic exists in the alternating tendency of the animal to "withdraw into" or to "emerge from" its shell, what Bachelard also terms "the dialectic of the hidden and the manifest" (110).

In the case of the latter, i.e., "images of 'emerging' " (109), there are the previously mentioned associations with birth, but additionally, a symbolism of "regeneration" (Eliade's word [173]) or of "resurrection" (employed by both Eliade [173] and Bachelard [114]), such that one might exclaim, with regard to the *coquille,* "Here then is a water phoenix" (BP 114). Incorporating intimations of both birth and death[10] and of a cycle from death to rebirth, the shell would seem to harbor "temporal explosions of being, vortices of being" (BP 110) and to disclose "an *élan vital* that turns" (BP 106).

In the case of the living mollusk's tendency to repair to its protective inner sanctum, the shell image evokes "reveries of refuge" (BP 107). Developing this theme, Bachelard relates the shell, as a symbol, to the nest (117), to the house (118), and to the fortress (125). For purposes of defense, the spiro-helical design embodies certain natural strategic advantages:

> As for spiraled shells, it is not "just for beauty, there is indeed something else . . . ; when [the fish] are attacked by their enemies on the threshold, withdrawing inside, they wind their way back [*ils se retirent en vironnant*], tracing a spiral path, and in this way their enemies cannot harm them." (124)

A safe and secure retreat, the shell also expresses "the isolation of being withdrawn into itself" (120), experiencing the quiet, languorous repose yearned for in dreams.

Anyone who has held a turbinate seashell up to his or her ear has discovered a magical property of this object: it seems to contain and amplify the very roar of the sea from which it issues.[11] And the cacophonous raging deep within the whorls immediately evokes the turbulent fury of wind and surf unleashed—whirling hurricanes, waterspouts, maelstroms. Here, we confront the spectrum of phenomena—also to a large extent sea-related—that are manifestations of the peculiarly dynamic form of the spiral helix signified by the English term "vortex" and the French *tourbillon.* Unlike the static, calcified swirl of the seashell, the vortex occurs in more rarefied—and hence more turbulent—fluid states, such as water and air or admixtures of the two. The word "turbulence" (from the Latin *turbare*)[12] betrays at once the kinetic nature of the phenomenon and the rotational disposition of the whirl.

The dynamic character of this symbol accounts, no doubt, for its association, from the outset, with power. It is primarily (though not exclusively) generative and creative forces that are signified in the ancient whirlpool image, which like the seashell, emerges first as a symbol of birth. In fact, the one is the other's dynamic double, and both are linked by their aquatic origin and congruent spiro-helical structures:

> The idea that life began in the first whirlpool, or came from the whorled shell which was connected with the whirlpool because of its spiral form, was perpetuated, and was as widespread as it was persistent. As in Gaelic, the shell and whirlpool were regarded as manifestations of the same life-giving force, which was symbolized by the imitative spiral in religious art. (MM 71)

Thus, in ancient Hindu mythology, the " 'embryo in which all the gods were collected' " issued from the "whirlpool of the primordial deep" (MM 70–71), indicating that the vortex image is also involved in the more elaborate birth symbolism of the creation myth.

Not surprisingly, though, it also carries destructive connotations, as is the case of the earliest vortex text my research has produced to date. The "Boulak Papyrus" has the distinction of recording perhaps the oldest vortical trope in literature, as well as one of the most curious:

> Keep thyself from the strange woman who is not known in her city. Look not upon her when she cometh and know her not. She is like a whirlpool in deep waters, the whirling vortex of which is not known. The woman whose husband is afar writeth unto thee daily. When none is there to see she standeth up and spreadeth her snare. Sin unto death is it to hearken thereto.[13]

Although Hobhouse is arguing that the position of women in ancient Egypt "was remarkably free" (187), he cites this passage as typical of the seemingly misogynist tone of early Egyptian literature, emphasizing that the warning is directed against either "the harlot" or "the adultress," not against women in general.

That the temptress should be compared to a whirlpool-vortex seems consistent with previously discussed characteristics of the symbol. The whirling fury of the natural phenomenon aptly mirrors both the passionate nature of the adulterous act and its disorienting, potentially destructive threat to the institution of marriage and to the social stability and tranquility that it presumably ensures. There is also the centripetal force that the whirlpool exerts upon objects coming within its influence, which parallels the irresistible attraction of temptation to those who come within its clutches. But why the whirlpool and not the whirlwind, which might also suggest these ideas? One notes, first of all, that the text specifies, "She is like a whirlpool in deep waters, the whirling vortex of which is not known." The fear of the unknown, like the fear of falling into a maelstrom, is obviously stressed, but less immediately apparent are the sexual overtones of penetration into "deep waters." In this sense, the whirlpool is a watery orifice with a spiro-helical structure, the dynamic equivalent of that other spiral helix, the turbinate seashell, which is associated, as we have seen, with the female genitals. Hence, the whirlpool-vortex proves to be an appropriate simile for the perils of adulterous seduction, radiating meanings on many levels in what is certainly one of the earliest vortex texts.

Another very ancient work that records examples of the *tourbillon* is the Bible, but it is the whirlwind, not the whirlpool, that figures prominently. Mackenzie has assembled many of the more important whirlwind symbols (76–77), although he does little more than list them. Apart from a recurring "stock" simile in Isaiah and Jeremiah, which depicts the whirling attack of chariots in battle,[14] the whirlwind is almost always associated, however indirectly or directly, with the deity and particularly with his awesome destructive powers. In some cases, this destructive force threatens to annihilate sinners or the wicked in general,[15] in other cases the enemies of God and of his chosen people ("the whirlwind shall scatter them")[16] or even the Jews themselves, especially for failing to "fear the Lord" and for idolatry.[17] The reference in Nahum (1:2–3) evokes vividly the anthropomorphic nature of the Old Testament deity:

> God is jealous, and the Lord revengeth;
> the Lord revengeth, and is furious;
> the Lord will take vengeance on his adversaries,
> and he reserveth wrath for his enemies.
> The Lord is slow to anger, and great in power,

> and will not at all acquit the wicked:
> the Lord hath his way in the whirlwind and in the storm,
> and his clouds are the dust of his feet.
>
> (B 637)

Quite unlike the perfect static entity unknowable to humans that emerges in Neoplatonist and Scholastic attempts to describe God from a New Testament perspective, the ancient conception of Yahweh, as anyone familiar with the Old Testament knows, suggests a supernatural being with a quasi-human temperament prone to paternalistic outbursts of ire, jealousy, and vindictiveness.

The whirlwind is thus an apt symbol of Yahweh's tempestuous nature and the destructive fury of his wrath. Nowhere is this identification more explicitly portrayed than in the Book of Job, where the Lord addresses Job "out of the whirlwind."[18] That God should manifest himself in this particular form seems consistent with his willingness in the story to visit destruction upon a devoted follower in a manner as ostensibly fickle and senseless as a tornado's sudden, devastating descent upon a tranquil village. But the vortical personification of God is fitting in other ways. II Kings 2:11 states that "Elijah went up by a whirlwind into heaven" (B 307), depicting, in effect, the conjunction of heaven and earth by means of the whirlwind. This is logical enough, since a whirlwind is an invisible force that spirals up toward or down from the sky, and Mackenzie documents "widespread lore regarding the association of supernatural beings with whirling gusts of wind" (74). Elijah is one of the few Old Testament figures to be "taken up" into the whirlwind—and the only one thus conveyed from earth to heaven—which suggests his subsumption directly into the deity, since the whirlwind is Yahweh's distinct emblem. This may account for the occasional confusion of Elijah with Christ or his traditional role "as one who will return as herald of the Messiah," a prophecy that is fulfilled, according to Christians, in the Transfiguration of the New Testament. Whatever the case, Elijah does acquire a supernatural stature beyond that of most other Old Testament figures, as a result of his advocacy of "the pure worship of Yahweh" and "the mystery with which his whole career was shrouded" (B 334). Still, his extraordinary induction into the celestial whirlwind must not be underestimated, as it imbued him with hypostatic divine and human resonances of the God-man he prefigured.

The cryptic, turbulent "visions of God," with which Ezekiel dramatically initiates the account of his call to prophecy (B 644–46), portray a vertiginous collage of fantastic image fragments, interrelating, through symbolic correspondence, a fiery whirlwind, four "four-faced winged creatures" (part man, part beast), spinning "wheels" (that also appear as "a wheel in the middle of a wheel" enclosing a spirit [1:16]) and other recondite whirling circular and tetradic synecdoches. Ezekiel's claim that "then the spirit took me up" (3:12) echoes Elijah's experience of rapture[19] but seems more a figurative demonstration of his privileged access to divine thought than a literal assimilation into godhead. Mackenzie detects in these visions allusions to the "gods of the cardinal points" (76); others might see prefigurations of the four evangelists or reflections of other tetradic paradigms.[20] From the perspective of the geometry of symbolism, the earthly rectilinear dimension (i.e., the cross of the four directions) is reconciled with the heavenly curvilinearity of the spinning wheels by means of the whirlwind's dynamic vorticity, portraying the mingling of heaven and earth that is the essence of the prophetic hermeneutic.

The vortex is thus a very powerful symbol of divine intrusion into the world and also a very appropriate one. Plato argues, in the *Timaeus,* that celestial circularity must necessarily be static to be perfect and unchanging, since motion implies change ("becoming") and "the eternal essence" ("being") is "immovably the same."[21] When extended into time, however, the divine circle metamorphoses into the whirlwind's spiral helix, which incorporates ingeniously the paradox of an immutable entity moving through change. Symbolically, the circular convolutions revolving progressively along a continuum now stand for assymmetrical cycles of time that turn about an unperturbed center—the deity—who is a "still point" of unchanging perfection from which all configurations in space-time emanate, as well as a "singularity" conducting to a mysterious other world toward which all things ultimately are drawn.

If in the Bible the whirlwind is almost always associated with the power of Yahweh, in the Homeric epics it rarely has anything to do with Zeus. In fact, the vortical image is all but absent in the *Iliad,* except for an occasional simile, like the description of the devastating prowess of Hektor, who "fought on like a whirlwind."[22] Achilles' two triple encirclings of the body of Patroklos with Hektor's corpse ignominiously in tow[23] are not obvious examples of vorticity, but they do seem to constitute a kind of ritualistic whirl. In Book IV of the *Odyssey,* Menelaos describes a similar triple encirclement, when he relates, in the presence of Helen (now retrieved) and Telemakhos, how the Greeks concealed in the horse were nearly exposed because of a curious action on Helen's part. Addressing her, he recalls,

> Three times you walked around it, patting it everywhere, and called by name the flower of our fighters, making your voice sound like their wives, calling.[24]

Only Odysseus's determined efforts to silence his comrades saved them from succumbing to Helen's sirenlike lure. Indeed, the cryptic ritual encirclings add to the mystery surrounding Helen, who like a sibyl—she is after all the daughter of Zeus—seems to be either casting a spell on the horse or undoing one in a "ceremony of riddance."[25] The incident also casts doubt on her true loyalties and intentions (did she really long to return to her homeland, as she claims?). What may appear to be an attempted betrayal of the Greeks is nonetheless portrayed by Menelaos as unwitting complicity, since she was doubtless "drawn by some superhuman power" (HO 61).

But the more explicit whirling images in the *Odyssey* involve the work's central protagonist, Odysseus. In Book I, Telemakhos states his belief that "the whirlwind got him, and no glory" (9). Although buffeted by turbulent storms throughout his wanderings, it is actually a whirlpool that nearly annihilates him during the exciting series of terrifying sea ordeals that dramatically conclude the first half of the *Odyssey* and for which the epic is renowned. The last and most treacherous of these is the perilous passage through the straits guarded by Skylla and Kharybdis, the former a dread six-headed serpent-monster, the latter a similarly rapacious whirlpool-vortex.

The episode, related in Book XII, comprises three separate events. In the first, Kirke prophesies the encounter and describes how Kharybdis swallows down the sea tide and sweeps to destruction everything caught up in it:

> Three times
> from dawn to dusk she spews it up
> and sucks it down again three times, a whirling
> maelstrom.
>
> (212)

Here, as before, the circular movement occurs in cycles of three, echoing the tripartite structure of the episode. But in the case of the vortex, the turning round involves spiro-helical expansions and contractions from center to periphery and vice versa, which constitute, in effect, three pairs of alternate centripetal (a systolic sucking-in) and centrifugal (a diastolic spewing-out) whirls. All told then, there are six individual stages of vorticity associated with Kharybdis's daily "routine," which correlate paradigmatically with Skylla's six-headed threat, and the peril is the same in both cases, i.e., the risk of being swallowed up and devoured alive, either whole in one gulp by the voracious vortex or piecemeal, six men at a time, by the ravenous serpent-beast.

For this reason, in the second event of the episode (the first actual encounter), Odysseus, heeding Kirke's advice, avoids Kharybdis altogether and takes his chances with Skylla, who claims her bloody forfeit in one of the most harrowing scenes in the epic and, for Odysseus himself, "far the worst I ever suffered" (218):

> we rowed into the strait—Skylla to port
> and on our starboard beam Kharybdis, dire
> gorge of the salt sea tide. By heaven! when she
> vomited, all the sea was like a cauldron
> seething over intense fire, when the mixture
> suddenly heaves and rises.
>
> The shot spume
> soared to the landside heights, and fell like rain.
>
> But when she swallowed the sea water down
> we saw the funnel of the maelstrom, heard
> the rock bellowing all around, and dark
> sand raged on the bottom far below.
> My men all blanched against the gloom, our eyes
> were fixed upon that yawning mouth in fear
> of being devoured.
>
> Then Skylla made her strike,
> whisking six of my best men from the ship.
>
> (217)

The narration is a masterpiece of irony, since the sailors, in fear of being swallowed up alive by one imminently menacing source, suffer the same fate unawares from a totally unexpected one—Odysseus had concealed the threat of Skylla to avoid panic—but the scene also represents, as far as the focus of this study is concerned, the first description from on the edge of a maelstrom. Also, on the symbolic level, it portrays the whirlpool-vortex not only as a redoubtable cataclysm but as the personification of a supernatural feminine force with its (her) own name and individual identity.

Although Odysseus and most of his crew survive the calamity, the passage through the straits represents the beginning of his solitary wanderings, as his comrades are ultimately destroyed by Zeus for having wantonly slaughtered Helios's cattle, in spite of oaths they swore to the contrary. Now, in the wake of the tempest, Odysseus, clinging to flotsam, drifts back through the straits, falling victim this time to Kharybdis's churning fury. In this third event of the episode, he manages to survive more by luck and divine intervention than by wile or wit, because he is able to grasp the branch of a jutting bough and avoid being sucked up into the whirl. As for Skylla, "Never could I have passed her / had not the Father of gods and men, this time, / kept me from her eyes" (224).

The episode of Odysseus's encounter with Skylla and Kharybdis is the most elaborate example of vortical symbolism in the ancient texts we have examined so far, and it typifies the Homeric conception of the cosmos in which natural and supernatural, human and superhuman forces are inextricably entwined. In the texts I now propose to study, the scope continues to be "cosmic," although the focus is no longer on a hero's struggle to overcome the forces of nature but on nature itself, what it is and how it came into being.

Cosmic Vortices of the Pre-Socratic Philosophers

2

Inasmuch as the episteme of the first philosophers comes after the Homeric age and before the period of Platonic rationalism, it is, as one might expect, transitional in the sense that it manifests characteristics of both epochs. From the former it often retains poetic symbolism as a privileged mode of signification. In anticipation of the latter it emphasizes reason in thinking and accords importance to the concept of the λόγος. In many ways, this period of intense philosophic inquiry is a matrix from which the entire Western tradition of discursive thinking issues. The approach to understanding *de rerum natura* already manifests the rudiments of a scientific method in its utilization of observation, hypothesis, and deduction to arrive at conclusions, and the focus is literally on φύσις, not the gods, as the source of truths and principles that can explain why things are the way they are.

The term "scientific" must not, however, be understood in the strict, modern, apodictic sense of the term, since there exists a concomitant tendency to speculate uninhibitedly about universal patterns and principles and to express these ideas in symbols. The Pre-Socratic thinkers are, to a great extent, poets, as Michel Serres avers, when in comparing the "ancient doctrine" of atomic physics to the "contemporary discovery" of it, he observes, "In the second case, it's a question of a science, that of Perrin, Bohr or Heisenberg; in the first, it's only a question of 'philosophy,' indeed of poetry."[1] This poetic quality is intensified, moreover, by the fragmentary nature of many of the extant texts and by the cryptic and at times aphoristic styles of the authors themselves.

As for the solutions proposed to the problems of "the nature of things," they usually center on one or more natural phenomena invested with both physical and symbolic properties, whereby the immanent "thing" represents an underlying causative principle or "idea."[2] So we discover, as early as Thales (c. 585 B.C.),[3] the identification of an element, ὕδωρ ("water"), which is more than a mere element, being the source of and thus prior to all things. For Anaximenes (c. 546/5), ἀήρ ("air") is preeminent and even divine. Concerning these early thinkers, Aristotle remarks,

> Of the first philosophers, then, most thought the principles which were of the nature of matter were the only principles of all things. That of which all things that are consist, the first from which they come to be, the last into which they are resolved (the substance remaining, but changing in its modifications), this they say is the element and this the principle of things.[4]

Aristotle's term for this "first-principle" is ἀρχή, but Kirk and Raven claim that he may have exaggerated its material nature and, in the case of Thales, underestimated "the cosmic importance of water" (89).

The same is true, they maintain, of his use of ἀρχή to describe Heraclitus's fundamental element, πῦρ ("fire"), which is probably related to divine αἰθήρ ("ether"), but which is certainly no longer "an originative stuff in the way that water or air was for Thales or Anaximenes" and functions more likely as "the continuing source of the natural processes" (KR 200). This interpretation derives from a particularly cryptic and no less provocative Heraclitean fragment, which is all the more interesting and relevant because it seems to be the first intimation of an epistemic vortex principle. For purposes of comparison, the translations of both Freeman and Diels and Kirk and Raven are included, respectively:

> 31. The changes of fire: first, sea; and of sea, half is earth and half fiery water-spout.... Earth is liquified into sea, and retains its measure according to the same Law as existed before it became earth.[5]

> 221. Fire's turnings: first sea, and of the sea the half is earth, the half 'burner' [i.e. lightning or fire]... (earth) is dispersed as sea, and is measured so as to form the same proportion as existed before it became earth.[6]

Kirk and Raven assert that "Vortices are not associated in our doxographical sources with anyone before Empedocles" (128) in an attempt to refute Aristotle's attribution of a vortex to Anaximander (a near-contemporary of Thales). Heraclitus, too, precedes Empedocles, and while the word "vortex" (δῖνος or δίνη) per se does not occur in his extant writings, the cosmic view conveyed in the fragment cited above, as Freeman and Diels translate it, does evoke the image of whirling fire and water. According to them, the word πρηστήρ[7] designates a "fiery water-spout"; for Kirk and Raven, the word is obviously problematic, and they offer " 'burner,' " "lightning," or "fire" as possible senses. If, for them, no whirl is implied in this term, they do read the opening two words of the fragment, πυρὸς τροπαί, to mean "Fire's turnings"; whereas Freeman and Diels see simply "the changes of fire," as if to transfer the connotation of rotation to the waterspout image.

Set in the context of Heraclitus's other pronouncements, this image does seem appropriate. To begin with, Heraclitus's philosophy is dialectical, i.e., concerned with the paradoxical unity (synthesis) of antithetical polar dualisms. Vorticity results from the meeting and mingling of contrary currents or from the slippage of fluid flows over one another as well as around obstacles and other contrary forces. And the notion of flux has long been associated with Heraclitus, due chiefly to his alleged statement that "In the same river, we both step and do not step, we are and we are not."[8] The vortical turbulence of fluid flow would thus seem a natural consequence of Heraclitus's notions of flux and the violent clashings of contraries.

Moreover, fire represents for Heraclitus "the controlling form of matter" (KR 200), effecting, as FD fragment 31 vividly illustrates, transformations between the three "world-masses," sea (water), fire, and earth.[9] It is therefore not only a fitting dynamic force for the philosopher's turbulent cosmogony, but like its nemesis, water, is conducive to the principles of vorticity.[10] Underlying fire, since it is "co-extensive with" it (KR 188), is the important Heraclitean concept of λόγος, which is translated by Kirk and Raven in the previously quoted fragment as "proportion" (elsewhere "measure" or "reckoning") and by Freeman and Diels as "Law." If then fire is vortical and if "the Logos is closely related to fire" (KR 201), is it not possible that λόγος, as "the structural plan of things" (KR 188) or the law of proportion in changes between opposites, embodied by fire, embodies as well the principles of vorticity? The fragmentary, cryptic nature of Heraclitus's extant writings makes it difficult to answer this question with certainty, but the evidence seems to indicate an affirmative response.

But it is precisely because Heraclitus's aphorisms, like oracular prose poems, are both fragmentary and cryptic that we must rely primarily upon the suggestiveness of symbolic multiple-sense in order to reconstruct the philosopher's cosmic view. In the case of Empedocles of Acragas (c. 444–41 B.C.),[11] the explanation of how things came to be what they are is wrought in verse,[12] and again, the extant texts are fragmentary and abstruse. As with Heraclitus, the basic "stuff" of the cosmos involves "elements" (four not three) and a "principle" (not a singular entity but a polar dualism).[13] Specifically, the universe is a Parmenidean sphere containing four co-eternal and distinct elements, fire, earth, air, and water—there is no void—that either come together under the influence of φιλότης ("Love," "Concord") or are scattered ("separated off") under the influence of Νεῖκος (variously, "Strife," "Hate," "Wrath," "Discord") as part of a cyclical "double process" (FD 53, fr. 17):

> A double tale will I tell: at one time it grew to be one only from many, at another it divided again to be many from one. There is a double coming into being of mortal things and a double passing away. One is brought about, and again destroyed, by the coming together of all things, the other grows up and is scattered as things are again divided. And these things never cease from continual shifting, at one time all coming together, through Love, into one, at another each borne apart from the others through Strife. (KR 326, fr. 423)

Kirk and Raven deem this cycle

> to have four stages, two polar stages represented by the rule of Love and the rule of Strife, and two transitional stages, one from the rule of Love towards the rule of Strife, and the other back again from the rule of Strife towards the rule of Love. (327)

This polar model, with movements between extremes, seems accurate enough as far as it goes, but Kirk and Raven's analysis of it ostensibly as four sequential stages does not account for the suggestion of a complex, synthetic simultaneity in the philosopher's insistence on a "double" process, i.e., "a double coming into being of mortal things and a double passing away."

Kirk and Raven also seem at a loss to explain the vortical movement implied by the use of ἑλίσσομαι, when the poet says of Aphrodite ("Love"): "Her does no mortal man know as she whirls around amid the others" (KR 328, fr. 424). Nor does their basic model come to grips with paradoxes of the δίνη ("vortex," translated "whirl" by Kirk and Raven) described explicitly in the following passage:

> When Strife had reached to the lowest depth of the whirl, and Love was in the middle of the eddy, under her do all these things come together so as to be one, not all at once, but congregating each from different directions at their will. And as they came together Strife began to move outwards to the circumference. Yet alternating with the things that were being mixed many other things remained unmixed, all that Strife, still aloft, retained; for not yet had it altogether retired from them, blamelessly, to the outermost boundaries of the circle, but while some parts of it had gone forth, some still remained within. And in proportion as it was ever running forth outwards, so a gentle immortal stream of blameless Love was ever coming in. (KR 346–47, fr. 464)

Kirk and Raven's reaction to the passage is one of consternation:

> But there is no denying that 464 is both vague in outline and obscure in detail. What, for instance, is the δίνη, 'whirl', of l. 4, and how did it arise? . . . And is its 'lowest depth', to which Strife is said in l. 3 to have fallen, the same as 'the outermost boundaries of the circle' in l. 10? It would seem that it must be so, but it is far from clear from Empedocles' own words. (347–48)[14]

Their bafflement derives from two basic questions: where does the vortex in this passage come from? and how is the presence of Strife both at the lowest point and at the circumference of the eddy possible?

To anyone familiar with the principles of vorticity, it seems obvious that the whirl results from the confrontation between the antinomies, Love and Strife, just as eddies are formed from the confluence of opposing currents. Aristotle, in fact, recognizes in Empedocles' process a vortical cosmogonic theory, which he describes (and criticizes) as follows:

> If, then, it is by constraint that the earth now keeps its place, the so-called 'whirling' movement by which its parts came together at the centre was also constrained. (The form of causation supposed they all borrow from observations of liquids and of air, in which the larger and heavier bodies always move to the centre of the whirl. This is thought by all those who try to generate the heavens to explain why the earth came together at the centre. They then seek a reason for its staying there; and some say, in the manner explained, that the reason is its size and flatness, others, with Empedocles, that the motions of the heavens, moving about it at a higher speed, prevent movement of the earth, as the water in a cup, when the cup is given a circular motion, though it is often underneath the bronze, is for this same reason prevented from moving with the downward movement which is natural to it.)[15]

Aristotle takes issue with the vortex theory not so much as a principle of causation but as an explanation—a specious one he claims—of how the earth stays in its central position. Beyond exposing possible inadequacies of the theory, his response is indicative of the tension, in this episteme, between the more rigorous "scientific" method espoused by Aristotle, concerned with the validity of the mechanics of the theory, and the "poetic" tradition which Empedocles is still a part of and which Aristotle ignores altogether. That Empedocles is proposing a symbol and not just a mechanism seems clear in his designation of the antithetical cosmic forces, not as "attraction" and "repulsion," but as Love and Strife, which are rich in connotative multiple-sense and capable of amplifying his vision beyond the scope of physics. Given this visionary quality, we should not be surprised if Empedocles' insight fails to withstand the minute scrutiny of an Aristotle.

Another commentator, W. B. Yeats, proves to be much more sensitive to the metaphorical possibilities of Empedocles' poetic text, and his analysis clarifies the structure of the symbol and reveals how it functions. It also accounts for the synthetic elements of a "double process" and explains the paradoxical simultaneity of Strife at both the periphery and the vortical core. This is because Yeats detects in Empedocles' system the intersection of two antithetical gyres:

If we think of the vortex attributed to Discord as formed by circles diminishing until they are nothing, and of the opposing sphere attributed to Concord as forming from itself an opposing vortex, the apex of each vortex in the middle of the other's base, we have the fundamental symbol of my instructors.

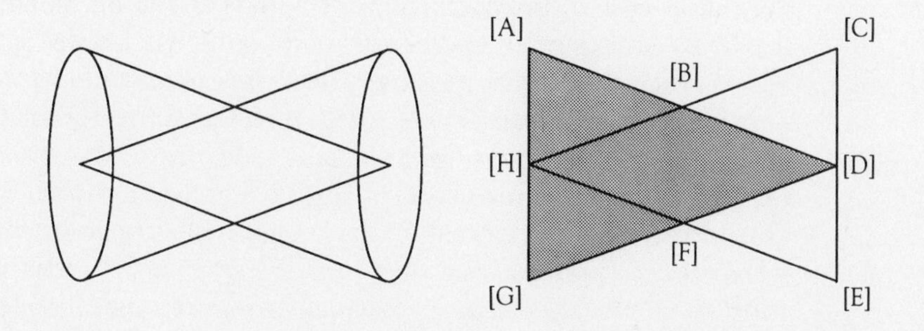

If I call the unshaded cone "Discord" and the other "Concord" and think of each as the bound of a gyre, I see that the gyre of "Concord" diminishes as that of "Discord" increases, and can imagine after that the gyre of "Concord" increasing while that of "Discord" diminishes, and so on, one gyre within the other always. Here the thought of Heraclitus dominates all: "Dying each other's life, living each other's death."[16]

Yeats's model not only illustrates a "double process," it depicts the concept of "one from many, many from one"[17] (in the preceding diagram, for example, AG ⇆ D or CE ⇆ H), and expresses, by the conjunction of equal circles at the median plane of intersection BF, the harmonious rule of Love at the center, suggested by the phrase, "Love in their midst, equal in length and breadth" (KR 328). Furthermore, the paradox of Strife at "the lowest depth of the whirl" and at "the outermost boundaries of the circle" is representable as either simultaneous (e.g., the relationship of H to AG) or sequential (e.g., H in relation to CE). In short, we have here a perfect example of dialectic (i.e., polarity, antithesis, and synthesis, simultaneously and/or in sequence) and, if we admit Yeats's assignation of contrary directions of rotation to the two gyres, an enantiomorph.

It is altogether appropriate, then, that Yeats should detect in Empedocles an echo of Heraclitus, as the quotation at the end of the previously quoted passage reveals, for the Empedoclean double process, because it is a dynamic, antagonistic synthesis of polar antinomies, not only offers a plausible explanation for change, but provides a mechanism capable of accounting for the "unity of contraries" that underlies so many of Heraclitus's paradoxical pronouncements.

With Anaxagoras of Clazomenae (c. 500/499–428/7 B.C.), this mechanism is further refined, insofar as it explains how the one becomes the many, but the mechanism itself, now clearly described as a singular rather than a dialectical entity, retains the polyvalent ambiguity of a symbol, due particularly to its origins in νοῦς (Mind). First of all, the "many from one" of Empedocles is restated by Anaxagoras as "the separated from the mixed," since the original unity—"all things together" (KR 368, frr. 495 and 496)—was a cohesion of all potential pluralities that were subjected to vortex action by νοῦς. This initiated a process of separating things according to the vortical principle of like with like, extending from heavier elements in the center through concentric circles of progressively lighter ones, with νοῦς itself, the lightest and the finest of all, apart and controlling the process:

> 503. All other things have a portion of everything, but Mind is infinite and self-ruled, and is mixed with nothing but is all alone by itself. . . . For it is the finest of all things and the purest, it has all knowledge about everything and the greatest power; and Mind controls all things, both the greater and the smaller, that have life. Mind controlled also the whole rotation, so that it began to rotate in the beginning. And it began to rotate first from a small area, but it now rotates over a wider and will rotate over a wider area still. And the things that are mingled and separated and divided off, all are known by Mind. And all things that were to be, all things that were but are not now, all things that are now or that shall be, Mind arranged them all, including this rotation in which are now rotating the stars, the sun and moon, the air and the aither that are being separated off. And the dense is separated off from the rare, the hot from the cold, the bright from the dark and the dry from the moist. But there are many portions of many things, and nothing is altogether separated off nor divided one from the other except Mind. (KR 372–73)

Anaxagoras's "whirl" is a subtle, equivocal, and cleverly devised entity.

On the one hand, it is a cosmogonic principle of organization, according to which matter has been arranged in space by Mind, resulting in a cosmos in which the earth rests at the center while the heavens whirl about it in ever-expanding concentric circles of rarer and rarer elements. On the other hand, the vortex, in addition to being spatial, is a temporal extension of Mind, describing a phenomenology of time in which past, present, and future are continuously widening spiro-helical cycles of centrifugal rotation, as denoted by the sequence of verb tenses in the following excerpt: "And it began [past] to rotate first from a small area, but it now rotates [present] over a wider and will rotate [future] over a wider area still."

As for Mind, the "intelligence" behind the vortical structuring of things, it is curiously reminiscent of Heraclitus's λόγος, which similarly "got things moving" in accordance with a law or "logic" of order and proportion, based on principles of vorticity, although, in that case, involving an apparent pyro-vortical process of tumultuous change. Unlike Empedocles' dialectic of Love and Strife, however, terms which might well be allegorical expressions of the contrary tendencies to adhere and to come apart, Heraclitus's λόγος and Anaxagoras's νοῦς, as the words themselves imply, possess distinctly intellectual characteristics of human consciousness, and this is all the more true for νοῦς, since as the above passage states, it is a "self" that "knows."

At this point, two problems that perplexed and even preoccupied the Pre-Socratic philosophers, eluding their attempts to resolve them convincingly and concretely, merit further exposition. The first, namely a "mechanical" explanation of how exactly the one became the many and why it did so, was partially solved by Anaxagoras. Inasmuch as his original "one" was a priori a heterogeneous conglomeration, the transition to multiplicity entailed merely an "ungluing." Why did this occur? Because Mind caused it to, for an unspecified reason, which brings us to the second problem. What, specifically, is this controlling principle (i.e., λόγος, φιλότης/Νεῖκος, νοῦς)? Is it a physical, spiritual, or hybrid entity? Is it divine? These questions are left disturbingly unresolved by Heraclitus, Empedocles, and Anaxagoras, although the latter's concept of "Mind" connotes a conscious, comprehending, deliberate motivator.[18]

In the next significant developmental phase of this episteme, what we now call atomism,[19] the two problems are, in effect, eliminated at the outset by radical challenges to the assumptions underlying them. By positing an infinite plurality at motion in a void, the dilemma of many from one no longer figured at all, and the ostensible abandonment of a controlling force did away with speculations concerning the identity of such a force. Still, the formation of the cosmos through vortex action and its spiro-helical configuration in space were retained and even emphasized by Leucippus and Democritus, and the word δῖνος ("vortex") now came to be associated openly and repeatedly with their cosmogony, as the following passage demonstrates, with its four mentions of δίνη and its use of the verb περιδινοῦμαι (which connotes a whirling motion):

> 562. Leucippus holds that the whole is infinite . . . part of it is full and part void. . . . Hence arise innumerable worlds, and are resolved again into these elements. The worlds come into being as follows: many bodies of all sorts of shapes move 'by abscission from the infinite' into a great void; they come together there and produce a single whirl, in which colliding with one another and revolving in all manner of ways, they begin to separate apart, like to like. But when their multitude prevents them from rotating any longer in equilibrium, those that are fine go out towards the surrounding void as if sifted, while the rest 'abide together' and, becoming entangled, unite their motions and make a first spherical structure. . . . This structure stands apart like a 'membrane' which contains in itself all kinds of bodies; and as they whirl around owing to the resistance of the middle, the surrounding membrane becomes thin, while contiguous atoms keep flowing together owing to contact with the whirl. So the earth came into being, the atoms that had been borne to the middle abiding together there. Again, the containing membrane is itself increased, owing to the attraction of bodies outside; as it moves around in the whirl it takes in anything it touches. Some of these bodies that get entangled form a structure that is at first moist and muddy, but as they revolve with the whirl of the whole they dry out and then ignite to form the substance of the heavenly bodies. (KR 409–10)

The atomists' debt to Anaxagoras is obvious in this passage, although a void now engulfs the vortex. Moreover, while Anaxagoras's whirl seems to rotate centrifugally, the vector of vorticity in the atomist δίνη, according to this important fragment at least, is ambiguous. As a cosmogonic formative system, it conforms to Anaxagoras's model, with "finer" bodies moving "out towards the surrounding void"; on the other hand, the membrane spinning within the whirl "is itself increased" because "it takes in anything it touches," which implies a centripetal movement and raises the possibility of a dialectical reversibility.

A further difference between the vortices of Anaxagoras and the atomists is the former's motivating force, νοῦς, which as I mentioned, the latter eliminate altogether. Their claim, though, that "worlds come into being" when "bodies of all sorts of shapes move 'by abscission from the infinite' into a great void" to produce a vortex poses problems of its own. What and where is the "infinite" in relation to the "void" that contains the whirl?[20] and why did the abscission into a whirl occur at all? The first question remains elusively unanswered, primarily because of confused semantics in the doxographical sources, while the second has elicited two conflicting responses. Aristotle states that the δῖνος arose spontaneously "from chance" (KR 413, fr. 567), which contradicts "the only extant saying of Leucippus himself" (413) that "Nothing occurs at random, but everything for a reason and by necessity" (413, fr. 568). This view is reinforced by Diogenes Laertius's assertion (concerning Democritus's thought) that "Everything happens according to necessity" (412, fr. 565), and Kirk and Raven attempt to dispel the apparent disparity by claiming, "For Aristotle they are chance events because they do not fulfil any final cause; but the atomists emphasized the other aspect of non-planned mechanical sequences, i.e., as necessity" (413). The terms "chance" and "necessity" are thus reconcilable if the random collision of atoms proceeds according to "*theoretically* determinable" (412) sequences of mechanical interaction, or in other words, "Every object, every event, is the result of a chain of collisions and reactions, each to the shape and particular motion of the atoms concerned" (413).

It is clear that the atomists' substitution of necessity for Anaxagoras's concept of Mind hardly simplified the problem of causation, nor did their emphasis upon mechanics reduce the vortex to a mere machine, as is evident in Diogenes Laertius's claim that, for Democritus, "the cause of the coming-into-being of all things is the whirl, which he calls necessity" (412, fr. 565). This confusion of the cause (necessity) and the mechanical means (the vortex) charges the atomist δῖνος with mystic symbolic resonances, and like its prototypes, it is as much a metaphor as a scientific cosmogonic principle. The same is true, as we shall see in the next chapter, of Plato, who develops further the concepts of vorticity and necessity, exploiting both the science of rational deduction and the allegory of myth in the elaboration of his own visionary cosmogony.

Plato's "Great Whorl"

3

Plato and Aristotle represent, in many ways, a culmination of the Pre-Socratic inquiry into the nature of things. While Aristotle's rigorous scientism tends to set him apart from his predecessors (including his mentor), Plato continues to share such affinities with them as the readiness to use symbols and images to illustrate abstract ideas. In the *Republic*, the "Allegory of the Metals," the "Allegory of the Cave" and the "Myth of Er"—to name a few such instances in just one of his works—prove that Plato is not exclusively a rationalist and dialectician. He is in every way a poet, a term unlikely ever to be applied to Aristotle. It is not surprising, then, that Aristotle finds so much to criticize in the ideas of those who came before him. In the case of their vortical theories, as we have already glimpsed, he shows little more than historical curiosity tinged with disdain and dismisses them as mechanically inadequate. For Plato, on the contrary, the vortex functions as much more than a vapid mechanism. It is a dazzling symbol pulsating with cosmic, mystic significance.

Although, in the *Phaedo*, Socrates expresses disillusionment with Anaxagoras's Mind and vortex (whose cause is exasperatingly unspecified), Plato offers, in the Myth of Er at the end of the *Republic*, an at once strange, striking, and eclectic vision of the universe, which is the product of a rational god-mind, not unlike Anaxagoras's νοῦς, and which the poet represents by an image evincing the structure of a concentrically circular free vortex. In order to demonstrate this last thesis, which may not be immediately apparent because of the difficulty of the texts involved, I shall attempt to elucidate the nature of the "Great Whorl" by adopting

Cornford's comparative technique,[1] although reversing the perspective, so that characteristics of the "World-Soul" in the *Timaeus* — Cornford's main concern — serve to illuminate facets of the symbol in the *Republic*.

The myth recounts the story of Er, a valiant soldier killed in battle, who died and came back to life with a report to the living of what awaits them after death. Plato's point is obviously an ethical one, aimed at convincing humanity that virtue is indeed rewarded and vice severely punished in the afterlife. Er's story opens with a description of two double processions of, respectively, righteous souls journeying "to the right and upwards" and returning "clean and pure," and unjust souls passing "to the left and downward" and coming back "full of squalor and dust."[2] Although Er is not himself punished or rewarded by the judges there, he is permitted to accompany the souls returning from heaven or hell on their way to ultimate reincarnation and reentry into the phenomenal world. And as all of them must, he too beholds a vision of the structure of the cosmos, which appears, at first from a distance,

> extended from above throughout the heaven and the earth, a straight light like a pillar, most nearly resembling the rainbow, but brighter and purer. To this they came after going forward a day's journey, and they saw there at the middle of the light the extremities of its fastenings stretched from heaven; for this light was the girdle of the heavens like the undergirders of triremes, holding together in like manner the entire revolving vault. And from the extremities was stretched the spindle of Necessity, through which all the orbits turned. Its staff and its hook were made of adamant, and the whorl of these and other kinds was commingled. (PRS 501 [616B–D])

I shall return to the important notion of "Necessity" when the significance of the symbol is considered as a whole. For the present, it is evident that Plato has borrowed from mythology the association of Fate with spinning, utilizing rather ingeniously the apparatus itself and, in particular, the whorl to depict the physical pattern of the universe.

> And the nature of the whorl was this. Its shape was that of those in our world, but from his description we must conceive it to be as if in one great whorl, hollow and scooped out, there lay enclosed, right through, another like it but smaller, fitting into it as boxes

> that fit into one another, and in like manner another, a third and a fourth, and four others, for there were eight of the whorls in all, lying within one another, showing their rims as circles from above and forming the continuous back of a single whorl about the shaft, which was driven home through the middle of the eighth. (PRS 501–3 [616D–E])

Cornford remarks that

> What the souls actually see in their vision is not the universe itself, but a model, a primitive orrery in a form roughly resembling a spindle, with its shaft round which at the lower end is fastened a solid hemispherical whorl. In the orrery the shaft represents the axis of the universe and the whorl consists of 8 hollow concentric hemispheres, fitted into one another 'like a nest of bowls,' and capable of moving separately. It is as if the upper halves of 8 concentric spheres had been cut away so that the internal 'works' might be seen. The rims of the bowls appear as forming a continuous flat surface; they represent the equator of the sphere of fixed stars and, inside that, the orbits of the 7 planets.[3]

The claim that the souls "actually see . . . a primitive orrery" seems dubious, although the spindle and whorl may be analogous to such a device. Cornford, in his attempt to clarify the image, reads into it elements that are not in the text. His statement that "It is as if the upper halves of 8 concentric spheres had been cut away so that the internal 'works' might be seen" is misleading if one infers that the symbol, as such, is really spherical and not hemispherical. That the perfect sphericity of the universe is not, indeed cannot be, visibly portrayed in toto seems to me an important signification of the symbol. After all, the purpose of the symbol is to convey to the living a palpable image of the cosmos, but because human beings are, as we know from the *Timaeus,* composite creatures—half mortal body, half immortal soul—they are equipped to understand only "half" of the system, that half which is phenomenally immanent, not the transcendent pure form of divine intelligence, the sphere itself, which alone is capable of fully fathoming its own essence.[4]

One must assume, then, that what Plato intended the symbol to be is what he actually depicts: a system of eight concentric hemispheres. And these are further described as follows:

> Now the first and outmost whorl had the broadest circular rim, that of the sixth was second, and the third was that of the fourth, and fourth was that of the eighth, fifth that of the seventh, sixth that of the fifth, seventh that of the third, eighth that of the second. And that of the greatest was spangled, that of the seventh brightest, that of the eighth took its color from the seventh, which shone upon it. The colors of the second and fifth were like one another and more yellow than the two former. The third had the whitest color, and the fourth was of a slightly ruddy hue; the sixth was second in whiteness. (PRS 503 [616E–617A])

Cornford, in his translation of this passage, interpolates the referent to which each of the preceding ordinal numbers corresponds—clarifying, though perhaps altering somewhat the text—and he demonstrates, in his reconstruction of the cosmology of the *Timaeus,* that the alignment of the Great Whorl is analogous to that of the set of concentric circles in the later symbol of the World-Soul. Moving centripetally from the periphery, the order of hemispheres (circles) is as follows:

1. Fixed Stars, broadest rim, spangled
2. Saturn, narrowest, like Mercury the yellowest
3. Jupiter, seventh broadest, the whitest
4. Mars, third broadest, slightly red
5. Mercury, sixth broadest, like Saturn the yellowest
6. Venus, second broadest, second whitest
7. Sun, fifth broadest, brightest
8. Moon, fourth broadest, color reflected from sun[5]

A curious anomaly becomes apparent once the circles are thus correctly identified. No mention is made of the earth, which one assumes is at the center, as it is in the World-Soul symbol of the *Timaeus,* although even there it is not a ninth orb,[6] as might be expected. Clearly, the hemispheres (circles) refer only to revolving bodies, and the earth rotates on the universe's central axis but does not revolve in an orbit. Neither the earth's position nor its behavior, however, is mentioned in Er's description of the cosmic motions:

> The staff turned as a whole in a circle with the same movement, but within the whole as it revolved the seven inner circles revolved

> gently in the opposite direction to the whole, and of these seven the eighth moved most swiftly, and next and together with one another the seventh, sixth, and fifth, and third in swiftness, as it appeared to them, moved the fourth with returns upon itself, and fourth the third and fifth the second. (PRS 503 [617A–B])

Set next to the description of the motions of the World-Soul, this passage illuminates and is illuminated by it, with the result that a fuller picture of Plato's cosmology emerges, as Cornford has shown, albeit from the opposite perspective of the *Timaeus.* The text in question describes how the Demiurge fashioned two strips of "soul stuff" into oppositely tilted circles that appear to intersect, like a horizontally skewed "X" on opposite sides of a diameter, although one circle is actually inside the other:

> He then comprehended them in the motion that is carried round uniformly in the same place, and made the one the outer, the other the inner circle. The outer movement he named the movement of the Same; the inner the movement of the Different. The movement of the Same he caused to revolve to the right by way of the side; the movement of the Different to the left by way of the diagonal. And he gave the supremacy to the revolution of the Same and uniform; for he left that single and undivided; but the inner revolution he split in six places into seven unequal circles, severally corresponding with the double and triple intervals, of each of which there were three. And he appointed that the circles should move in opposite senses to one another; while in speed three should be similar, but the other four should differ in speed from one another and from the three, though moving according to ratio. (PTC 73–74 [36C–D])

Cornford has commented at length upon the significance of this text and has found it compatible with the cosmology in the Myth of Er, except for the inclining of the seven inner circles at an angle to the outermost, but the World-Soul image provides more information concerning the contrary motions of the inner and outer rings. The latter is right-handed and called "the Same"; the former is left-handed and termed "the Different." The opposition between the purely spiritual celestial motion and that of the mixed corporeal and spiritual circles of the Different is one of direction, in which the heavenly is "to the right" and everything below "to the left," parallel to the directions, in the Myth of Er, of the just souls ("upwards to the right") and the unjust ("downward to the left"). In short, sunwise and witherwise motion acquire distinctly moral connotations in Plato's thought, and the opposition of Same and Different implies a conflictual cosmos that recalls Empedocles' dialectic of Love and Strife, although, for Plato, the Same is supreme and pervades the entire universe, whereas the Different is confined to the seven inner rings.

A close reading of both texts reveals that the motions of the inner bodies are not uniformly to the left but appear at times to move in contrary directions to one another. Cornford discusses this dilemma in great detail. The problem is that the occasionally "erratic" movement of certain planets, notably that of Mars, which to Er "appeared" to move "with returns upon itself"—what Cornford translates "with a counter-revolution" (PTC 88)—suggests that the inner directions are reversible, contradicting Plato's original differentiation of the opposite movements of the Different and the Same. Cornford, in his solution, emphasizes the word "appeared" and demonstrates, via the analogy of a "moving staircase," that the discrepancy is an illusion caused by relative motions. The entire universe revolves in the direction of the Same, but the seven inner circles, moving oppositely in the direction of the Different at varying velocities, counteract this movement. In the case of some planets, resistance to the Same is weaker than that of the others, and while the weaker ones continue to be swept along in the direction of the Different, they "appear" to move contrarily in relation to the other planets.

Two consequences of this orbital dialectic deserve comment. The first is the "spiral twist" (PTC 112–14) detectable in planetary retrogradations, which is particularly apparent in the annual double spiral traced by the sun as it fluctuates seasonally between the solstices.[7] Although these irregularities are due to optical effects caused by the earth's revolution about the sun in a heliocentric system, Plato's turbulent vectorial conflict is an original, if in the end unsatisfactory, solution. A second consequence of these opposing motions is that the earth, which, Plato says in the *Timaeus*, "winds round the axis that stretches right through" (PTC 120 [40C]) the universe (the "shaft" of the Spindle of Necessity), rotates with the axis in the direction of the Same but simultaneously exerts an equal and opposite motion of its own in the direction of the Different, such that the contrary motions cancel each other out, and the earth, as it were, stands still in the center: " 'The complete immobility of the terrestrial globe is, consequently, the result of two forces of rotation, whose physical effects annul one another, and one of which belongs to her intelligent soul.' "[8]

To complete now this reconstruction of Plato's complex symbol, it remains only to establish the order of velocities indicated in the previously quoted passages concerning cosmic motions. In this regard, the Myth of Er is more specific than the *Timaeus,* although both generally concur on the matter. The swiftest is the smallest ring and closest to the center, that of the moon; the slowest is the largest, that of Saturn, the farthest planet from the center. The intervening motions form a gradual progression between these extremes, with the sun, Venus, and Mercury, second in swiftness, moving in unison, Mars the third swiftest, Jupiter fourth, and Saturn, as we have seen, the least swift. It is thus evident that the velocities vary in a manner inversely proportional to the distance from the central axis, which when taken together with the pattern of concentric circles corresponds to the properties (described in the Appendix) of a "free circular vortex." There can be no doubt, then, that Plato, in the symbol of the Great Whorl, presents the stunning image of a vortex whirling within an oppositely revolving, all-encompassing hemisphere, as in a cup.

But Er's description of the vision ends as it began, not with "mechanics," but with an allusion to the mythological metaphor of "the spinning of Fate":

> And the spindle turned on the knees of Necessity, and up above on each of the rims of the circles a Siren stood, borne around in its revolution and uttering one sound, one note, and from all the eight there was the concord of a single harmony. And there were three others who sat round about at equal intervals, each one on her throne, the Fates, daughters of Necessity, clad in white vestments with filleted heads, Lachesis, and Clotho, and Atropos, who sang in unison with the music of the Sirens, Lachesis singing the things that were, Clotho the things that are, and Atropos the things that are to be. And Clotho with the touch of her right hand helped to turn the outer circumference of the spindle, pausing from time to time. Atropos with her left hand in like manner helped to turn the inner circles, and Lachesis alternately with either hand lent a hand to each. (PRS 503–5 [617B–D])

A question immediately arises concerning the meaning of "Necessity" in this passage. It was an important, if somewhat hazy, principle in the philosophies of Parmenides and the atomists, but it appears to be even more cryptic in Platonic thought, as Cornford demonstrates in his analysis of the opposition between "Reason" (νοῦς) and "Necessity" (ἀνάγκη) that the philosopher mentions in a celebrated passage of the *Timaeus:*

> For the generation of this universe was a mixed result of the combination of Necessity and Reason. Reason overruled Necessity by persuading her to guide the greatest part of the things that become towards what is best; in that way and on that principle this universe was fashioned in the beginning by the victory of reasonable persuasion over Necessity. If, then, we are really to tell how it came into being on this principle, we must bring in also the Errant Cause—in what manner its nature is to cause motion. (PTC 160 [47E–48B])

That Necessity is capable of being "persuaded" has led some critics to infer an element of "reason" within Necessity (the inexorable workings of natural laws?), but the reference to an "Errant Cause," at the end of the passage, has suggested to others that Necessity stands for contingency, chance, accident, or spontaneous occurrence (Cornford seems to side with the latter interpretation). Still others propose such definitions as "the indeterminate," "unexplainable fact," "destiny." The evidence in the *Timaeus* is decidedly ambiguous, perhaps because Necessity is a complex of some or all of these factors, as the symbol of the spindle in the *Republic* would imply. In fact, what seems to elude theoretical exegesis in the *Timaeus* is more accessible, more readily grasped aesthetically in the great symbol of the earlier work.

The last previous quotation from the Myth of Er places the spindle, i.e., the cosmos, "on the knees of Necessity," suggesting that she is the ultimate final cause, a base, an underlying framework, an axiomatic bottom line beyond which further inquiry is futile. Considering that Plato posits three givens, that is, "three distinct things" that existed "even before the Heaven came into being," namely "Being, Space, Becoming" (PTC 197 [52D]), are not these the "necessary conditions" upon which the cosmos is constructed? In other words, for the universe to exist at all, it "necessarily" had to be a synthesis of Being ("always the same, uncreated and indestructible, never receiving anything into itself from without, nor itself going out to any other, but invisible and imperceptible by any sense, and of which the contemplation is granted to intelligence only"), Becoming ("perceived by sense, created, always in motion, becoming in place and again vanishing out of place, which is apprehended by opinion and sense") and Space ("eternal, and admits not of destruction and provides a home for all created things").[9]

In the image of the spindle, Space is obviously the context in which the symbol "takes place." Being, which for Plato is divine and synonymous with reason, harmony, "the good, the true, and the beautiful," is symbolized by the divinities present (Necessity, her three daughters, the Fates, and the Sirens) as well as by the hemisphere of the Same, which Clotho, who speaks of "the things that are"—the divine eternal present—turns in the celestial right-handed direction. But the more cogent representation of Being is, perhaps, the absent (although implied) second hemisphere that we must imagine a perfect, though invisible, pure form, in comparison to which the imperfect system of concentric planetary orbits, with their retrogradations and other deviations, is but a shadowy mirror image.

Becoming, then, and especially the processes of change (generation and decay, creation and destruction, life and death) are symbolized by this visible, material extension of the static divine circle dynamically (i.e., temporally) into space; and the image of the vortical whorl effectively illustrates the paradoxical, mixed nature of this dimension, which is prone to turbulent "spiral twists" or other such divagations but which functions in an orderly manner according to the vortical principles of "like with like" and "inverse circular velocities." And it is Atropos who turns the inner hemispheres in the left-handed, earthwise direction of the Different, she who sings of "the things that are to be," signifying the progressive thrust toward an unknown future through stages of change that are characteristic of the visible world. The concept of Becoming in time is further implied in the process of spinning, as a metaphor for the human life cycle that the Fates, in their role of "the Moirae" (in French *les Moires*), personify: "Clotho (the Spinner, who spins the thread of life, Lachesis (Disposer of Lots), who determines its length, and Atropos (Inevitable), who cuts it off."[10]

Here, in the figure of Lachesis, we discover the important element of chance, for it is from her "lap" that the souls, in the Myth of Er, receive the "lots and patterns of lives" (PRS 841 [617D]) from which they must choose. But Plato, in his symbol, assigns her an additional function.

As the one who, "alternately with either hand," spins both the outer and inner rims, she depicts the harmonious dialectic that binds heaven and earth, which the Sirens, each imparting a note of the Pythagorean scale, literally intone.[11] The Fates further enhance the "concord" by singing "in unison with the music of the Sirens." Plato thus insists that, just as the heavenly motion of the Same is superior to and pervades the motion of the Different, divine Reason "got the better of necessity" (PTJ 30 [48A])—not by means of total subjugation but through compromise ("persuasion")—infused the universe, and produced the harmonious cosmic synthesis that the symbol so vividly depicts.

Given the rich complexity and paradigmatic depth we have just seen, it is hard to understand Cornford's disparaging dismissal of the Great Whorl symbol as "primitive" (PTC 75) vis-à-vis that of the World-Soul in the *Timaeus*. The latter with, as Cornford diagrams it (PTC 73), dual rings tilted asymmetrically is hardly a compelling figure and illustrates little more than the opposing circuits of the Same and the Different. The Spindle of Necessity, on the other hand, while standing for this (minus the tilt), resonates simultaneously with eclectic multiple meanings that include Homeric, mythological, astrological, cosmological, theological, ontological, and poetic-aesthetic elements, to name some of the levels. In this sense, the Great Whorl represents a culmination, an apex, in the development of an epistemic vortex symbol, which first emerged in the cosmic philosophy of the Pre-Socratics and next to which the subsequent whirls of Epicurean cosmogony, that I now propose to examine, may seem elusive and somewhat more mechanical.

Epicurean Whirlings and Lucretius's Turbulent Flux

4

That vorticity figures at all in the thought of Epicurus[1] is denied by Mackenzie, who, relying on a secondary source, states that the philosopher expressly "rejected the vortex" (108). The difficulty arises from the only passage in his cosmogonic writings that mentions a "whirl,"[2] including an apparent critique of the atomist δῖνος, followed by a reference to δινήσεις ("whirlings") in the exposition of his own view:

> For it is not merely necessary for a gathering of atoms to take place, nor indeed for a whirl and nothing more to be set in motion, as is supposed, by necessity, in an empty space in which it is possible for a world to come into being, nor can the world go on increasing until it collides with another world, as one of the so-called physical philosophers says. For this is a contradiction of phenomena.
>
> Sun and moon and the other stars were not created by themselves and subsequently taken in by the world, but were fashioned in it from the first and gradually grew in size by the aggregations and whirlings of bodies of minute parts, either windy or fiery or both; for this is what our sensation suggests.[3]

Bailey, in his notes, offers the term "vortices" as a synonym for the "whirlings" of the second paragraph and points out that they refer "to the independent rotation of the revolving nuclei, not to the δῖνος which causes the movement of the heavenly bodies through the sky" (286).

The question then is whether Epicurus subscribes to the "cosmic whirl" formation theory or rejects it, as some have interpreted the criticism in the first paragraph to mean. A close scrutiny of the text reveals that he is not refuting the atomist whirl per se, but the idea that it came into existence "by necessity." A strict materialist, Epicurus seems troubled by the teleological implications of such a mysterious force, for he emphatically denies the operation of intelligence or design in the generation of the cosmos.

Ironically, the atomists' attribution of the whirl to "necessity" was a reaction against Anaxagoras's νοῦς, but their solution, as we have seen, raised more questions than it answered, particularly with regard to the meaning of necessity as "chance" (Aristotle's αὐτόματον)[4] or "the inevitable functioning of innate mechanical laws." Kirk and Raven, we noted, do not find the two definitions mutually exclusive and seem to concur with Bailey's assessment of the atomist view: "the formation of the whirl is the outcome of natural causes, the shape, size, motions and collisions of the atoms, but it is accidental, entirely undetermined either by purpose or design."[5] Epicurus's own concept of the "swerve"—we shall presently examine Lucretius's eloquent account of it—is not altogether incompatible with the notion of necessity, as formulated above, and his criticism of the atomists' cosmogony must not be interpreted as a blanket rejection of the vortex theory.

In fact, Bailey concludes that "Epicurus had in general accepted the conception of 'the whirl,' "[6] if for no other reason than its being consistent with the way naturally occurring eddies appear to operate, and Epicurus insists, of course, upon the reliability and veracity of undistorted sense impressions.[7] Vortex action does figure in his explanations of such meteorological phenomena as thunder, thunderbolts, and cyclonic storms. Still, the vortical image seems more an implied, or perhaps even assumed, aspect of his cosmic theories, and it certainly lacks the symbolic multivocity of other examples we have seen.

The case of Lucretius[8] would seem similarly ambiguous concerning the cosmic whirl, to read two critics on the subject. Mackenzie claims the poet "says nothing about a vortex" (108). Michel Serres, on the other hand, asserts that "for Lucretius as for us, the universe is a global vortex of local vortices. Just like his poem" (157). Mackenzie's denial may result from the subtle way in which the *turbo* is woven into the texture of the full text and from misleading cues, such as his rejection, in Book I, of the Stoic argument that "all things tend / Toward a center."[9] Like Epicurus, from whom it is assumed most of the basic philosophy of the poem derives, Lucretius seems intent on avoiding the "necessary" connotations of the Democritean δῖνος, and characteristically, he offers it as one of several plausible explanations of the structure of the cosmos.

> Democritus, that hero of the mind,
> Whose judgment we revere, may well have come
> As close as anyone, his argument
> Being that when the stars are nearer earth,
> They are bound to move more slowly, cannot race
> With sky's full whirlwind sweep, whose force dies down,
> Whose impetus slows, in lower atmosphere.
> And so the zodiac signs behind the sun
> Catch up and overtake him, since his course
> Is much lower. The same way with the moon
> Or even more so; farther from the sky,
> Nearer the earth, by that much less her speed,
> By that much less her power to match the pace
> Of the zodiac signs; she moves below the sun
> With weaker whirling, and it seems almost
> That she is going backward, as those signs
> Come sweeping by her.
>
> (178)

It is not clear from the extant atomist fragments whether Democritus conceived of a cosmic whirl with streamlines increasing in velocity proportionally to the radius, as suggested above, since the opposite is the case in naturally occurring vortices. Here, Lucretius's reliance upon sense perception—the apparent speeds of celestial bodies relative to one another—without any recourse to analogical deduction, has led him into error.

There are, though, other less puzzling aspects of his cosmology that are easily recognized features of atomist vortex theory, such as the "separating off" of "like elements joined with like" (172) and the division of "parts arranged in order" (173), as well as a concentration of "heavier" elements—weight, for the Epicureans, is emphasized over size—at the middle of the whirl. The lighter particles are accordingly "squeezed out" to form the more fluid watery, airy, and fiery materials of progressively peripheral layers. And the poet's portrayal of the fiery ether floating unperturbedly above the maelstroms of lower airy regions not only recalls previous distinctions between the steady, unchanging circle of fixed stars and the erratic "deviations" of lower spheres but is a passage rife with Lucretian vortical vocabulary, as is evident in the original Latin:

> nec liquidum corpus turbantibus aeris auris commiscet;
> sinit haec violentis omnia verti turbinibus, sinit
> incertis turbare procellis, ipse suos ignis certo fert
> impete labens. (V: 502–5)[10]

But rather than saying that *the* whirl or even *a* whirl is at work in the philosopher's cosmogony, it is perhaps best to speak of many vortices coming into and going out of existence in ever-changing complexes of colliding atoms and swarming fluid flows that pervade the entire poem.

The first evocation of turbulence comes early in Book I, just after the poet has announced his atomic theory:

> The wind
> Beats ocean with its violence, overwhelms
> Great ships, sends the clouds flying, or at times
> Sweeps over land with a tornado's fury,
> Strewing the plains with trees, and beating mountains
> With forest-shattering blasts; its roaring howls
> Aloud and wild, and even its mutter threatens.
> Surely, most surely, the winds are unseen bodies,
> Sweepers of earth and sea and sky, and whirlers
> Of sudden hurricane. They flow, they flood,
> They breed destruction just the way a river
> Of gentle nature swells to a great deluge
> By the increase of rainfall from the mountains,
> Commingling in ruin broken brush and trees.
> Strong bridges cannot hold the sudden fury
> Of water coming on; the river, darkened
> By the great rain, dashes against the piles
> With mighty force, and with a mighty sound
> Roars on, destroying; under its current it rolls
> Tremendous rocks; it sweeps away whatever
> Resists its surge. So the wind's blast must also
> Be a strong river, a fall of devastation
> Wherever it goes, shoving some things before it,

> Attacking over and over, in eddy and whirl,
> Having its way, seizing and carrying things.
>
> (28)

The comparison of the flux of atoms to invisible tempestuous winds and a rushing river—the principal vortex words are *turbine* and *vertice*—leaves little doubt that Lucretius's world is one in which awesome vortical forces batter things about, dashing them ultimately to destruction, in a manner consistent with his repeated admonition that his is a stern message, "a bitter medicine" that he will attempt, through poetry, to mask "with honey's sweetness, honey's golden flavor" (46).

A parallel image opens Book II:

> How sweet it is, when whirlwinds roil great ocean,
> To watch, from land, the danger of another,
> Not that to see some other person suffer
> Brings great enjoyment, but the sweetness lies
> In watching evils you yourself are free from.
>
> (52)

The phrase *turbantibus aequora ventis,* which recurs in Serres's study as a kind of leitmotif and a slogan, as it were, of Lucretius's cosmic view, dramatizes the destructive nature of turbulence and the threat it poses to people's safety. The pleasure referred to, as the poet points out, is not a cynical, sadistic joy at another's misfortune, but the "avoidance of pain" that motivates Epicurean behavior. The allusion, later in Book II, to "That maelstrom of confusing otherness" (67)—Humphries' rendering of *materiae tanto in pelago turbaque aliena* (l. 550)—evokes again an image of raging seas and, by analogy, of violent clashings between random streams of atoms, but the emphasis here is upon chaotic disorder, since the word *turba,* as Serres points out, "designates a multitude, a large population, confusion and tumult. It is disorder." *Turbo,* by contrast, "is a round form in movement like a whip-top or spinning-top, a cone which turns or a turbulent spiral" (38); in other words, a system, an order that forms, however fortuitously or momentarily, in the midst of turbulent disorder. The allusion, somewhat earlier in the same book, to whorled conches—*concharumque* (l. 374)—strewn along the seashore is, in this sense, a symbolic synecdoche of the orderly *turbo* that emerges from the chaotic *turba* of the sea.

In Book IV, the vortical theme is implicit in an image offered by the poet to support his case that optical illusions are not the fault of the senses, which do not distort, but of reason, which misinterprets the sense impressions:

> Children, dizzy
> After they stop spinning themselves around,
> Think that the rooms revolve, the pillars whirl,
> And even ceilings threaten to fall down.
> (130)

Although this occurs as one image among many and is not therefore, in and of itself, a particularly salient one, it does demonstrate the variety and the ubiquity of the vortex symbol throughout *De Rerum Natura*. It also embodies an important corollary correspondence that resonates in the symbol's intricate scheme of paradigmatic associations. Vertigo is the feeling of dizzy disorientation that results from spinning or being spun around, as in a vortex,[11] and it stands, here as in other authors, for the dynamic alternation between states of order and disorder that the vortex effects.

We have already examined the turbulent elements in Book V of Lucretius's cosmogony, although two prefigurative, anticipatory whirling images were not mentioned, a reference to "doom and destruction" wrought "with almost whirlwind violence" (170)—*violento turbine* (l. 368)—and the image of the "sun's horses" that "whirled" (*deturbavit*) Phaeton across the sky. While incidental in themselves, they advance and intensify the vortical theme that pervades the texture of the poem and culminates in Book VI, which, as Serres puts it, "manifests in concrete examples the theory of physics" (107).[12]

The great sweep of meteorological phenomena that the poet-philosopher describes in the last book of *De Rerum Natura* is at once an extension of his assault against religious superstition (which often deems these forces to be of divine origin) by offering rational explanations, and a final, dramatic evocation of the vortex pattern into which atoms naturally "fall." In this respect, Lucretius surpasses earlier Epicurean efforts in the "Letter to Pythocles" by presenting conjecture after conjecture in an attempt to embrace all the possible causes. Strodach, in his notes to Humphries' translation, suggests the scope of his endeavor: "here, for example, Lucretius sets down nine separate explanations of thunder, and later, four for lightning, four for thunderbolts, three for earthquakes, and so on" (253). In the case of these and other phenomena, a vortical cause is always at least one of the theories offered or implied. Some of the significant vocabulary of turbulence—admittedly taken out of context—is as follows: in the passage on thunder, *turbine versanti* (l. 126); on lightning, *ventus . . . versatus* (l. 175) and *volvenda* (l. 179); on thunderbolts, *vertex versatur* (l. 277); on waterspouts, *presteras* and *prestera* (ll. 424 and 445),[13] *spirantibus* (l. 428), *versabundus enim turbo* (l. 438), *vertex* (l. 444) and *turbinis* (l. 447); on earthquakes, *versabunda* (l. 582); and, finally, on volcanic eruptions, *volvit* (l. 691) and *turbida* (l. 693). Given such a piling up of vortical images, it is hard to understand how so astute a scholar as Mackenzie can profess that the vortex was "ignored by Lucretius" (108).

Incontrovertibly, then, vortices do abound in Lucretius's text, but it now remains to determine whence and wherefore. The explanation offered is essentially that of Epicurus, although Lucretius's poetic rendition in Book II is actually a primary source for reconstructing Epicurus's theory. Given atoms moving in a void with a natural tendency downward, they will fall in parallel straight lines, never touching, their rate of speed constant, due to the fact that varying weights do not affect velocity in a void. Then, suddenly, spontaneously, an atom may "swerve"[14] slightly from its course, the deviation "no more than minimal" (58)[15] and occurring "at no fixed time, at no fixed place whatever" (60).[16] As a result of this "tiny swerve,"[17] a concatenation of atomic collisions creates the turbulence from which the forms and compounds of the visible world emerge.

Michel Serres has studied this cosmogonic model in depth, and his many insights are valuable and noteworthy. In the first place,

> Lucretius describes two chaoses: the stream-chaos, laminar flow of elements, parallel flux in the void, delineating a filamentary space; and the cloud-chaos, disorderly mass, fluctuating, Brownian, of dissimilarities and oppositions." (42)

Although the original "laminar flow" may seem at first sight to be "an orderly scheme" (37), it is, in fact, a chaos of unchanging conformity, "a statics of movement" in which, paradoxically, "the flux of atoms is inert" (60). It is also the product of the "laws of fate" (59)—*fati foedera* (II, l. 254), elsewhere *foedera naturae*—i.e., inherent properties of the atoms, such as indivisibility and motion in a downward trajectory, that are "imprinted in the core of things" (185) by means of "encoding." But to avoid just such a static determinism, Lucretius insists, the *clinamen* (swerve) must occur to make possible both change and free will, which the senses tell us do exist.

Thus, the swerve, which Serres describes variously as a "logical absurdity" (9), "a differential, a fluxion" (216), and "the smallest condition conceivable for the first formation of a turbulence" (13), represents "a breach of the fundamental laws of cause and effect, for it is the assertion of a force for which no cause can be given and no explanation offered" (BG 320). Chance, then, according to the Epicureans, is the irrational force that disturbs the monotonous parallel filaments of the stream-chaos and produces the turbulent phenomenal world. This turbulence is of two types, as we have seen: the disordered ferment of colliding motes, i.e., the *turba* of the cloud-chaos, and the ordered vortex system of the *turbo* (*tourbillon*), the latter, as it were, superimposed upon the former: "Figure in relief against a background, the vortex appears on the chaos, and the *turbo* on the *turba*" (SN 41).

And yet the system, as Serres sees it, is dynamic and dialectical, as well as cyclical. At the poles, extremes of chaos—the homogeneous disorder of stream-chaos or the conflictual heterogeneity of cloud-chaos—in the middle, the orderly, if temporary, equilibrium of the *turbo*, which underlies all phenomena ("Every object coming into being is first a vortex, as is the world in its entirety" [SN 64]). Moreover, once the irrational *clinamen* disturbs (*dis-turbare*) the monotony of the "laminar flow," the formation of a *tourbillon* inevitably ensues, claims Serres, in accordance with the mathematical demonstrations of Archimedes.

> Here is then, again, the model. Given a shower of parallel streams, where the laminar flow glides, at any point, that is, at random, a swerve occurs, a very small angle. A vortex forms, from that point, immediately. (20)

Scrutinized closely, this vortex is a marvel of diverse, complex, and paradoxical qualities. As embodied in a spinning top, for example, which appears at once to move and stand still, this "unity of opposites" would seem to defy the Platonic-Socratic dictum that "the same thing will never do or suffer opposites in the same respect in relation to the same thing and at the same time" (PRS I, 381–83 [436B]). The crucial phrase for Plato is "in the same respect," for the top is stationary with respect to its axis while in motion with respect to its circumference. Serres, who points specifically to this example, argues against Plato's narrow interpretation, and he identifies additional paradoxical elements that inhere in the top and in the *tourbillon* generally:

> Release then this toy, describe, as Plato has done, what is happening. It is moving, there's no doubting that, yet it is stable. It is even at rest on its tip or its pole, all the more so if the movement is rapid. All children know this. But this stasis is more paradoxical still. The top can change its place, by translation, without ever ceasing to keep its stability. Again, it can do it on condition it turns very fast. Better still, its axis can tilt, incline, without imperiling too much the overall movement. It can still balance, by nutation, oscillation about an intermediate situation. This very ancient and puerile machine is a marvelous teacher.
>
> It reunites, first of all, all the known conceivable movements of the time: rotation, translation, falling, inclination and balancing. An integral, additive, surcharged, and yet simple, model. But secondly, and above all, it associates in a unique and easy experiment phenomena judged or presumed contradictory. It is moving and it is at rest, it turns and does not move, it teeters and it is stable. Simplicity of a complexity, first, and additive machine; synthesis of contradictions, more than any other thing. (39)

Serres observes further that, being at one and the same time a stationary and a rotating object that describes a circumference, "the spinning-top is a *circum-stance*" (40).

Given the dialectical-synthetic nature of the *turbo,* it is not surprising that the poem itself vacillates between antinomies: Venus and Mars (reminiscent of Empedocles' Love and Strife), "isonomy and entropy," "equilibrium and drift," "ataraxia and turbulent flow." But because the natural tendency is downward, "a declination," the *tourbillon,* from the moment of generation, is declining toward its ultimate destruction and dissipation into the chaos from which it emerged and in which it is immersed:

> Things are born from the swerve. They constitute themselves from this difference from equilibrium or from this minimal angle. From this small solid cone named *turbo.* From the moment of their

> coming into being, or their inchoate formation on the sheet of the fall, they are committed to return to the cataract. In the process of being born, destined to die, mortal nature. They are adrift on the *talweg* [steep slope] of the descent. And declination is drift, *talweg*, descent. On the whole, things decline. This theorem signifies that they must come undone, finished [finite], in their elements, at the end of their temporary existence, but also that this existence occurs only by decline. Decline is time. Its length or its interval, its beginning and its end. To be born is to decline. But also to exist, and as well, to die. (114)

We may conclude, then, that the Epicurean view, although optimistic on the general, universal level, since new swerves perpetually create new phenomena, is, in the case of specific, individual entities, tragic, for the aggregates of atoms, whether simple or as complex as a human being or an entire world, are subject to progressive degradation and ultimate annihilation. Thus the trajectory of the poem (which, for Serres, is "written in the form of a vortex" [172]) from the initial evocation of "spring" and Venerean harmony to the final, precipitous plunge into the squalid decay of plague-infested Athens at the mercy of Mars's destructive onslaught. Thus, too, the tragic *élan* of human being:

> I am myself a swerve, and my soul is declining, my body global, open, adrift. It glides irreversibly, on the slope. Who am I? A vortex. A dissipation that is coming undone. . . . Vessel adrift, under the angle of the helm, such am I. (50)

In the final analysis, the Epicurean-Lucretian solution to the troublesome Democritean attribution of the cause of the δῖνος/*turbo* to "necessity" involves an abandonment altogether of the term and a strict separation of its constituent aleatory and deterministic elements. There are *foedera naturae* that govern the interactions of atoms, such as the properties of the motes themselves. But the eventual, if unpredictable, deviation of the *clinamen* threatens at all times spontaneously to give the lie to mechanical laws, to disrupt their blind chain reactions, to contradict the logic of their *dénouement*, to overthrow an old and set in motion a new order, until it too, like a teetering top, its energy spent, runs down, wobbling out of control, succumbing finally to inevitable undoing, decline, entropy.

Having arrived, with Lucretius, at the end of this inquiry into the vortical symbolism of ancient texts, it seems only fitting to conclude the first part of this study with a brief general assessment of the findings. As might be expected, the destructive potential of the vortex, as embodied in whirlwinds or maelstroms, is a primary connotation of the symbol, especially in the earlier texts, where it is portrayed, variously, as an adulterous threat to social order ("Boulak Papyrus"), as a means of exacting divine revenge for faithlessness and corruption (the Bible), and as a formidable obstacle to a wily hero's homecoming (Homer's *Odyssey*). Most remarkable, perhaps, is the identification of the whirlwind with Yahweh, either as a personification of him and of his omnipotence or as a means of subsumption into divinity and, thus, of the mingling of the earthly and celestial realms.

But by far the most persistent and consistent development of the symbol is the image of a cosmic whirl that recurs throughout the period of classical antiquity that spans, roughly, the first six centuries B.C. Now the creative, constructive potential of the symbol is highlighted along with or even at the expense of its destructive characteristics, and instead of being identified with a stern, paternalistic, personal deity, as in the Old Testament, the source of its power is, rather, an impersonal force or principle with few anthropomorphic qualities, although the very intelligence of this entity is analogous to human consciousness. Heraclitus's *logos,* as we have seen, is a principle of change, i.e., of "fluctuating" states of creation and destruction, that appears to realize these transformations via the primary element fire as embodied in a vortical *prester,* or "fiery waterspout." In Empedocles, there are dual forces, Love and Strife, effecting a dialectical "double process" of turbulent shifts between harmony and discord. Anaxagoras's "Mind" is, once again, a singular force and is perhaps the most anthropomorphic of classical causes, since it "knowingly" brought the many from the one by means of a whirl. With atomism, the very notion of an intelligent creative and directive force is eschewed, and the mechanical properties of the *dinos* are underscored, although its identification ambiguously with "necessity" seems to have provoked the complicated solutions of Plato and the Epicureans. For Plato, the cosmos is an intricate cooperative interaction or "pact"[18] between chance, mechanical principles, a divine craftsman (the Demiurge), and the "necessary" preconditions of Space, Being, and Becoming, so cleverly synthesized in the vortical symbol of the Great Whorl. But Plato's belief in a divine order underlying the imperfect phenomenal world is totally abandoned by Epicurus and Lucretius, who designate chance alone the causeless first cause that engenders, by means of the swerve, a turbulent cosmos of both parallel and colliding atomic streams, a pervasive double chaos, from which stable eddies of order emerge ephemerally and into which they inevitably dissipate, only to be regenerated spontaneously by another swerve, at another time, elsewhere in an infinite void.

And indeed the vortex image proves to be rich in symbolic polyvalence, as well as a synthesis of complex and even contradictory properties: at once illustrative of the one and the many, static repetition and progression, ascendancy and decline, birth and death, creation and destruction, nascent order ("like with like," gradations of velocity) and moribund disintegration, stillness and movement (translation, rotation, oscillation, inclination, falling, hovering, balance), centripetality and centrifugality, efficiency and entropy, equilibrium and disruption of the status quo, ataraxia and flux—in short, a perfect symbol of the "circum-stance" in which human being discovers itself.

Nevertheless, despite its evolution into so rich and complex a symbol by the end of the classical age, it is not surprising that the vortex was eclipsed along with so many other ancient ideas during the ensuing thirteen centuries that witnessed the efflorescence of Christianity and now usher us to our next author. An image associated with "pagan" philosophies, espousing such concepts as necessity, chance, chaos, and turbulent flux, would almost certainly be targeted for oblivion by a culture committed to the theology of an omnipotent, omniscient, purposeful, and benevolent deity. It is all the more ironic, then, that the vortex image, like a hidden text uncovered beneath the inscribed surface of a palimpsest, should reemerge as a quintessential symbol of the late Medieval Christian vision in the work of Dante, whose *Comedy* incorporates some of the most brilliant and ingenious examples of vortical symbolism.

Part II

Visionary Breakthrough

Dante's Vortical Triptych

5

Just as ancient philosophers found in the whirl a phenomenon capable of explaining and illustrating the nature of the cosmos as they conceived it, Dante discovered in its circular, spiral, and vortical properties a reservoir of symbolic possibilities with which to fashion his mystical vision of the afterlife. These properties abound in the poem, not only in the very architecture of each of the three nether worlds but, recalling Lucretius's technique in *De Rerum Natura,* as recurring synecdoches that reflect and intensify the fundamental forms in images and episodes the poet describes during the course of his pilgrimage.

If we begin by scrutinizing the structure of hell, it is evident that the pit formed by the progressively narrowing concentric circles, translated along a central axis in the manner of a telescope, is an appropriate symbol of not only the theological concept of "evil as privation" (hence, a hole or abyss) but also the hierarchical degrees of sin that "increase in gravity" proportionally with the increasing gravitational pull of Dis at bottom center. The downward plunge is also an apt portrayal of the Fall and of the *inclination* toward evil resulting from original sin. But what exactly is the "vast funnel-shaped cavity,"[1] which Botticelli has rendered so vividly and meticulously in vertical cross section?

The word Dante uses most frequently to describe the physical configuration is *giro* (from the Greek *gyros,* circle or circulation), a word seemingly as ambiguous in Italian as its English equivalent "gyre," although the idea of circular motion is the common thread linking the various meanings of the noun *giro*[2] and the verb *girare.*[3] That an essentially circular form is implied is indicated by Dante's own reference to the whole structure of hell as *Li empi giri* (*Inf.* X:4).[4] Singleton's translation, "the impious circles" (1:1:X:99),[5] unlike Sayers's rendering, "the impious gyres" (1:X:128),[6] clarifies the circular nuance and shows that the singular *giro* refers to only one wheel-like level at a time. This distinction is an important, if subtle, one, since it is often assumed that Dante's hell has a spiral or spiro-helical shape, which is a possible connotation of the singular term "gyre" in English, but not, apparently, of *giro* as used by the poet in the *Comedy.*

Although not spiro-helical, the structure of hell is nonetheless vortical, displaying the concentric streamlines of a "free circular vortex" and conforming quite closely to this model with one major exception. Whereas in a naturally occurring free vortex velocity varies inversely to the radius, motion from rim to center of the Inferno decreases proportionally with the radius. This is not a hard and fast rule, and the activities of the damned vary widely from circle to circle with a confused lack of order that is part of Dante's depiction of evil. Nevertheless, there is a significant contrast between the frenetic turbulence near the mouth of hell—the futile madly chasing a whirling standard or the lustful buffeted aimlessly on a howling wind in Canto II—and the lugubrious, constricting punishments of Malbowges—the simoniac popes stuffed headfirst into fiery pits, the barrators wallowing in boiling pitch, and the traitors either half-buried, like the giants and Dis himself, or imprisoned up to the neck or completely engulfed by the ice. This apparent devolution from a dynamic, if erratic, activity at the periphery to the stagnant statis of Cocytus at the core manifests the structure of a free vortex in which the velocities are backward. And such reversal is consistent with Dante's deliberate attempt to portray hell as backward and unnatural in every way.

Another instance of this topsy-turvy structuring has been studied by Freccero. The question concerns the direction of circumvolution in hell. Is it leftward and thus "sinister," as one might suspect? Freccero emphasizes the difficulty of determining direction on a circuit when only left-handedness or right-handedness is indicated: "Briefly, to know which is the 'right' is not enough; we must be told which is front and which is back, or which is up and which is down" (FG 170). The paradox of circular direction (see Appendix) is readily apparent if we imagine an observer following the movement of the second hand of a clock he or she is facing. From the 9 o'clock to the 3 o'clock position the hand moves from left to right, imitating the apparent motion of the sun as viewed from the northern hemisphere, whereas the same second hand, continuing past the 3 o'clock to the 9 o'clock position shifts directions and moves from right to left. Since, however, we do not observe that half of the sun's apparent course through the heaven, it is clear that we determine direction of circumvolution from the path traced by the upper arc, not the lower one.

Let us imagine, further, Dante entering the circles of hell, facing the pit, and announcing a turn to the left. Relative to the full circumference of the pit, he would find himself at the bottom arc. Thus his initial leftward impulse would, after passing into the upper arc, shift to the right, and his direction for the circuit as a whole we would describe as clockwise or to the right. Unfortunately, the poet is not all that accommodating or consistent; in fact, the directional notations in the *Inferno* are sparse, confusing, and equivocal, as Freccero demonstrates, but he concludes ultimately that "from incidental details in the poem we know that the pilgrim's path is generally clockwise throughout hell" (FG 170).

If the descent is "always to the left" (DSI XIV:149), as Virgil exclaims to Dante in Canto XIV, and hence clockwise (right-handed), the movement in hell is in the natural, sunwise direction, not widdershins, i.e., the unlucky counterclockwise sense one might expect. Freccero's comments concerning this dilemma are interesting and revealing. Briefly, Dante's cosmology is essentially Aristotelian. In his treatise, *De Caelo,* Aristotle concludes that "South is 'up' in the cosmos and North 'down' " (FG 170), which means that east, as it is when facing the sun in the southern hemisphere, is "to the right" and west "to the left." These are the "true" directions that are inverted in the northern hemisphere. I detect in this opposition of hemispheres a distinction between a false "fallen" world mired in evil (the northern) and a "true" earthly paradise free of sin (the southern), similar, but not identical, to the visible and (implied) invisible hemispheres of Plato's Great Whorl. Northerners—the only inhabitants of the world known to Dante—are, consequently, upside-down and at the bottom of the sphere. Although the pilgrim may have the impression, from the northern perspective, that his left-to-right clockwise direction in hell is "sunwise," this is an illusion, since the "true" solar direction, i.e., that of the southern perspective, is right-to-left or counterclockwise. In this sense, Dante's clockwise descent is widdershins.

Contrarily, it could be argued that his descent is "truly" an ascent or "upwards" (FG 170) and his direction in hell correspondingly reversed (counterclockwise or in the true sunwise direction). Given the relativity of perspective and the way it can affect meaning, I think it important to read the symbols first as Dante presents them. His entrance into hell occurs in the north, so from this perspective (*the* perspective of the *Inferno*), it is a clockwise descent, a fall into the abyss of evil. This does not preclude comparing northerly direction of the sun to its true southerly counterpart and judging the former witherwise *by comparison*. But we must not accept the southern perspective as the given perspective until the traveler actually arrives there, as he will in the *Purgatorio*. And I would suggest one last point concerning the important question of perspective. If, in the *Inferno*, direction and orientation seem hopelessly confused and contradictory, if things in the northern hemisphere are not as they appear and it is easy for one to lose one's way, as Dante does at the opening of the poem, isn't this precisely his point about the imperfect phenomenal world, where humans make false choices based on illusions and plunge ever so readily into sin?

The confused, illusory, backward, and unnatural nature of hell is thus intentionally incorporated into its structure, as is the eternal, tautological monotony of the punishments meted out to the damned, which the independent circular gyrations of each level vividly depict. In a manner inversely analogous to the souls in heaven, who, as we shall see, are fixed in circles of eternal bliss, each sad sinner in hell is condemned to repeat, in a vicious circle that whirls on forever, that evil choice of a lesser love over the greater divine love, the loss of which is irrevocable, as he now knows all too well and has an eternity to regret. In this respect, the negative connotations of Dante's infernal vortex are consistent with its traditional portrayal as a violently destructive, deadly force, although in keeping with "paradoxes of the vortical core" (see Appendix), Satan stands a paralyzed, impotent, pathetic lord of a frozen, eerily still, and silent wasteland, "in the eye of the hurricane," while all about him reel ever-widening gyres of chaos and cacophony.

The association of vortical turbulence with sin is not Dante's own idea—notwithstanding his original elaboration of it—but an Augustinian convention, embedded in the Catholic tradition from which the poet of the *Comedy* draws his inspiration. In the *Confessions*, it is primarily to sins of concupiscence that Augustine is referring when he speaks of "the whirlpool of debasement"[7] that characterized his early childhood and when he later condemns "the whirlpool of sin" (43) into which lust and fornication had plunged him. His friend Alypius's blood-lust at the amphitheater games, he also notes, was the result of being "caught in the whirl of easy morals at Carthage" (120).

The phrase "caught in the whirl of easy morals" has a distinctly Dantesque ring to it but in a more general and pervasive sense than Augustine's, as should be apparent in the study of gyro-vortical images that I now propose to undertake. The importance of this image is immediately evident in Dante's comment, upon passing through the gate of hell into the "vestibule of the futile":

> Here sighs, laments, and loud wailings were resounding through the starless air, so that at first they made me weep. Strange tongues, horrible outcries, utterances of woe, accents of anger, voices shrill and faint, and the beating of hands among them, were making a tumult that swirls [*s'aggira*] unceasingly in that dark and timeless air, like sand when a whirlwind blows [*turbo spira*]. (DSI III:27)[8]

The feeling of vertigo that results at the end of that canto in Dante's first full faint is already apparent in his initial reaction to the scene ("And I, my head circles with error . . .)," as if he himself were caught up in the dizzying spins of the "fence-sitters," who, prodded by stinging wasps, futilely chase a whirling—*girando* (III:53)—banner. The "weather-cock mind" (DSA 89) of those whose standards shifted with every prevailing current or mode is evidence not merely of "easy morals" but, in effect, of no ethical principles whatsoever.

In Canto V, the vortical image recurs in two scenes. First, when Dante and Virgil come upon Minos, the poet, playing on the word *volte* (V:11 and 15), describes how the ferocious judge signals to each soul the circle appropriate to its trespass by coiling his tail about him as many "times" as the level to which the soul thence is "whirled." The gyrations of the sinner's plunge are thus cleverly reflected and anticipated in the paradigm of the monster's coiled tail (a helical configuration that will be addressed presently along with other such tropes). In the second scene, Dante beholds the punishment of the lustful in the second circle, as they are tossed about pell-mell in a black wind:

> I came into a place mute of all light, which bellows like the sea in tempest when it is assailed by warring winds. The hellish hurricane, never resting, sweeps along the spirits with its rapine; whirling [*voltando*] and smiting, it torments them. (DSI V:49)

Like Augustine's whirlpool, Dante's whirlwind image associates the passionate excess of unchecked carnal lust with the destructive turbulence of a vortex and is a striking example of how the poet "makes the punishment fit the crime."

The lot of the homosexuals, in the seventh circle, is comparably chaotic, according to Sayers, who observes that "their perpetual fruitless running forms a parallel, on a lower level, to the aimless drifting of the Lustful in Canto V" (165). And when, to keep pace with Dante, the three homosexual shades of Canto XVI form a circling wheel, enabling them to keep moving as they must while simultaneously staying in place relative to Dante, the vortical whirl is once again evoked, which embodies paradoxically properties of both motion and stasis, as Serres noted in the "circumstance" of a spinning top.

Perhaps the most famous vortical image in this volume is the whirlwind in which Ulysses perished. The story of his last voyage is certainly a curious, if original, feature. Readers of the *Odyssey* no doubt wonder what source provided the model for this self-obsessed deceiver, condemned with the other counsellors of fraud to the eighth circle. Of course, the poet's own championing of the Trojan-Roman cause would have rendered this Greek hero an enemy, and his crafty involvement in the treachery of the Trojan Horse had already contributed to his reputation among later Greeks as a cunning and ruthless trickster.[9] But the tale of his last voyage, in which the wanderer willfully forsakes father, wife, and son and the duty and piety owed them, is a complete reversal of Homer's story, which seems to have been the traditional version in the Middle Ages. As Odysseus is rewarded for his faithfulness at the end of the *Odyssey* with glory and the prospect of a peaceful old age, Dante's Ulysses is, inversely, brought to ruin prematurely by hubris and unbridled ambition (a fitting Christian punishment of pride). Also, like Dante himself in the dark wood at the foot of Mount Purgatory, who was unable (because unworthy) to attain the mountain and fell into hell's vortex, so Ulysses and his crew fell victim to a *turbo* in sight of a mountain that seemed to promise refuge and a safe landing:

> There appeared to us a mountain dark in the distance, and to me it seemed the highest I had ever seen. We rejoiced, but soon our joy was turned to grief, for from the new land a whirlwind [*turbo*] rose and struck the forepart of the ship. Three times it whirled her round [*il fé girar*] with all the waters, and the fourth time it lifted

> the stern aloft and plunged the prow below, as pleased Another, till the sea closed over us. (DSI XXVI:279–81)[10]

It is clear here, as elsewhere in the *Inferno,* that the vortex represents an image of doom. Like Yahweh concealed in the whirlwind, it testifies to the destruction that divine justice will visit upon the wicked. It remains, however, to discuss what role, if any, the helical and spiro-helical configurations play in the scheme of hell, since the structure of hell itself, as we have seen, is concentrically circular, not spiralic. Once again, the image of Minos signaling his judgment by the convolutions of his tail is relevant. The coils are connected by the single strand of the tail itself, unlike the discrete circles of the vortex. Symbolically, the helix implies a progression from ring to ring that the circular vortex does not; and just how this image figures in the overall vortical scheme must now be determined.

As each soul is whirled by Minos to its destination, a progressive helical plunge must certainly ensue, comparable to the leap from shore of some of the shades ferried by Charon across Acheron that the poet compares in the previous canto to leaves falling in autumn (III:112–17). Some leaves because of their size and shape spin downward in a helical path. Others are tossed to and fro in a zigzag. This simile is a clue to the pattern of continuous descent that Dante and Virgil alone are capable of tracing, for they are anomalies destined to pass through and out of hell as no others among the damned may ever hope to do. Nevertheless, though a specifically spiro-helical descent is clearly implied,[11] it is not consistently realized, and their actual path, as diagrammed by Sayers (84, 122, 180, 264), traces, variously like the leaves, the *talweg* of a meander or (leftward) spiro-helical encirclings. Let us consider the situation more closely.

Frequently, in the *Commedia,* absolute conditions are implied that would be impossible for the poet to represent literally. I just cited, for example, Dante's description of the second circle as "a place mute of all light," and yet he is able to discern the shades of Paolo and Francesca. Later, deep in the ninth circle of the traitors, after having, for some "unexplainable reason," kicked Bocca degli Abati in the face, Dante responds to his curses by violently wrenching a few tufts of hair from his scalp, despite the fact that the victim, entombed helplessly to the neck in ice, is presumably an impalpable shade lacking material substance. So it is with the spiro-helical descent. It would be unfeasible for Dante and Virgil, in the space of the one hundred cantos allotted, to make a full revolution of each gyre. The eighth circle alone comprises ten trenches.

When they arrive at circle seven, it is clear they have not yet done so, since Virgil remarks to Dante: "You know that the place is circular; and though you have come far, always to the left in descending to the bottom, you have not yet turned through the whole circle" (DSI XIV:149). Thus, the leftward meander mimicks the helical alternation back and forth, while shortening the course, and its ostensibly erratic zigzag (like the flight of the starlings in Canto V, *di qua, di la*) reflects the pilgrims' confused, crooked journey into the depths of sin.

That the helical encirclings are nonetheless implied is clear when, at the edge of the immense precipice separating the seventh and eighth circles, a network of helical images is evoked that underscores the full structural pattern of Dante's infernal descent. He and Virgil have just left the three Florentine homosexuals who whirled alongside them tracing a series of helical circuits in an effort to keep abreast, while their heads, riveted on Dante, twisted in the opposite direction (a not very subtle negative Christian depiction of homosexuality as unnatural and contrary). Now, at the edge of the cliff, Dante at Virgil's bidding loosens the cord girt round his waist, coils and knots it, and passes it to his guide, who flings it forthwith into the abyss. This cryptic gesture, on the one hand, serves as a "strange signal" (DSI XVI:171) to the monster Geryon below, beckoning him to rise, but the curious coiling and knotting also prefigure the coiled serpentine form and brilliant "knots" of color that characterize the foul brute.

The principal coiled image, however, is the helical path the travelers take, astride the beast, in response to Virgil's command: "Geryon, move on now; let your circles be wide, and your descending slow" (DSI XVII:179).[12] The fantastic flight is indeed a breathtaking image in and of itself, but Dante further intensifies the symbolism through paradigmatic correspondence, first, in a vortical reference to "the whirlpool"—*il gorgo* (XVII:118)—roaring horribly below them as they soar and, second, in a striking helical bird simile:

> As the falcon that has been long on the wing—that, without seeing lure or bird, makes the falconer cry, "Ah, ah, you're coming down!"—descends weary, with many a wheeling [*per cento rote*], to where it set out swiftly, and alights disdainful and sullen, far from its master: so, at the very foot of the jagged rock, did Geryon set us down at the bottom, and, disburdened of our persons, vanished like an arrow from the string. (DSI XVII:181)

This is one of many references to falcons, hawks, or other birds of prey in the *Commedia,* but it is unique for the aspect of falcon behavior it portrays, since the bird's dramatic "stoop"—a vertical "dive-bomb" to catch its prey unawares on the wing—is the behavior more typically depicted.[13] Here, the falcon's "weary" circlings denote not so much the predatory instinct of hell's creatures—the bird is actually abandoning the hunt—but rather the resentment of lost freedom and enslavement to Satan that Geryon displays, comparable to the "disdain" for taming that a once-wild falcon might exhibit by "sullenly" and rebelliously refusing the lure.

It is clear, then, not only in the now implicit, now explicit pattern of descent itself, but in corresponding metaphoric figures reflecting the form, that Dante's voyage through hell is helical (or, more precisely, spiro-helical) from the cone's rim to Lake Cocytus at dead center. In purgatory, we discover many analogous details of structure and motion, although in this case, they are likely to relate inversely to constituent elements in hell. Compared to hell, the spiral, helical, and vortical imagery of purgatory is simpler and inheres more in the general design than in details of the successive cornices.

When studied closely, the structure of purgatory reveals itself to be an enantiomorph of hell, turned inside-out. It would be hard to imagine a more dialectical antinomy. Instead of a funnel-shaped void depressed into matter, a solid, substantial cone rises up into the air and light forming a seven-tiered mountain. As in hell, the levels consist of progressively narrowing circular gyres, but in place of hollow trenches along the pit's concave inner side, steplike cornices telescope up the cone's convex surface, like a wedding cake.

The idea of such a sacred mountain is a very ancient one, even though Dante's conception is original and unique. Pyramids are obvious prototypes, and the step pyramids are not only among the oldest found, but the basic form or a variation of it is as persistent as it is universal. The comparatively recent pyramids of the Mayan and Aztec civilizations in the New World are invariably storeyed structures. Middle Eastern ziggurats, by comparison, share the rectangular base of the pyramidal forms, but the tiers slope up at a constant angle about the core, forming a continuous, quadrangular spiro-helical path to the top.[14] In England, a spiral route winds about the hill on which Glastonbury Tor stands (see PM, pl. 41). In this case, the mountain is a natural feature that has been shaped somewhat roughly into a symbolically tiered sacred site. Perhaps the closest analog to Dante's purgatory is the upper part of the shrine at Borobudur, Java (an immense edifice comprising three circular terraces superimposed upon five that are square).[15] Even the Great Stupa on top seems to parallel the earthly paradise as a place of spiritual enlightenment to which the pilgrim must progressively aspire, although Dante's symbolic journey involves circular revolutions from base to summit.

But what precisely does the poet's purgatorial trek entail? Once again the comparison with hell is instructive. In both places, Dante uncannily respects the law of gravity that stipulates increased gravitational pull (i.e., heaviness) as one approaches the center of the earth, although the symbolism is reversed, since in hell one succumbs easily to one's "fall" toward the center, whereas, in purgatory, one must vigorously resist the downward gravitational pull in order to move upward and away. Here, as before, gravity serves to illustrate the Christian concept of original sin, due to which humans have an almost irresistible inclination to evil that requires rigor and ardor to "surmount."

Accordingly, just as the graver sins are punished at the bottom of the pit, so are they purged at the foot of the mountain, but the bottom in hell is the goal of the voyagers, while in purgatory their point of departure. The movement toward the goal, moreover, is inverse in each case, decelerating during the progressive descent toward Dis, as has been shown, and accelerating with the ascent of Beatrice. On the second terrace of ante-purgatory, Belacqua huddles languorously in his fetal cocoon, waiting to move up, although he, like everyone in purgatory, will eventually—for him it may well take millennia—attain the summit. Somewhat higher up, on the first cornice, the proud do move along, but lugubriously, each one straining under the weight of an enormous stone that serves to humble his overly exalted self-image. By contrast, on the seventh cornice, the lustful run swiftly, breathlessly, heterosexuals in one direction, homosexuals contrarily (again reflecting the negative stereotype), although the groups pause briefly to embrace each other in a gesture of harmonious cooperation, which is the very spirit of purgatory's society.

Activity does then intensify with the ascent, along with ease of movement, which increases in a manner inversely proportional to the gravitational pull, or in Virgil's words: "This mountain is such that ever at the beginning below it is toilsome, but the higher one goes the less it wearies" (DSI 2:1:IV:41). In effect, the velocities here are consistent with those of a naturally occurring free circular vortex, just as those of hell are "unnaturally" reversed. But purgatory is a middle ground, and the vorticity is, consequently, only half realized, for in the dark of night, all is silent and still, whereas during the day, the energizing light and motivating divine love whirl the souls around and up. To sum up, then, purgatory is an inverted, inside-out, diurnally pulsating vortex, i.e., the sacred mountain that reverses the damage done by sin, quite the opposite of the treacherous infernal maelstrom that lures sinners to their destruction, sucking them up in eternal death. And the trajectory traced by the penitent from base to summit is (as in hell from rim to core) spiro-helical.[16]

The direction of circulation, however, is the opposite of hell's, since Dante and Virgil move up and "to the right." According to Freccero, "there seems to be no doubt that the ascent of the Mount of Purgatory is counterclockwise" (168). If we imagine Dante, at a given moment, facing the mountain in the 6 o'clock position, *a destra* would imply a movement toward 3 o'clock. At this point, the direction shifts from right to left, moving toward 9 o'clock, and since direction of revolution is determined by this half of the circuit, it follows that Dante is turning in a counterclockwise sense.

One's first reaction might be to wonder how negative widdershins could be symbolically appropriate for purgatory, which exists to "set things right"? Again, the question of perspective is all important. From the northern perspective of the audience for whom Dante wrote, the direction is witherwise and perhaps even appropriately so, since purgatory represents for them a last chance to "undo" the damage caused by evil, to disabuse themselves of wicked habits, and to get rid of all taint and stain due to sin. In this sense, the leftward encirclings of the mountain denote what Mackenzie would call "a ceremony of riddance" (see Appendix) analogous to the concept of purgation. Freccero isolates, moreover, a passage attributed to St. Augustine in which he claims to have been "unwound" from his previous "occupation in evils," and Freccero concludes, "Before the soul can make progress, the twisted course of the will must first be unwound" (178).

But the predominant perspective must be, it seems to me, that of the pilgrim of the story, who finds himself, at this point, in the southern hemisphere, where all is reversed, and the sun, as noted earlier, moves in the "true" right-to-left direction. The counterclockwise motion up the mountain, from this ideal perspective, gyrates in harmony with the sun, conveying the penitent from the winding hollow, where he has been righted ("converted" [FG 178]), through each cornice's cleansing circuit to the pinnacle of virtue in the earthly paradise and from there whirling into the heavenly spheres.

If the voyage through hell and purgatory furnishes the poet with an image of his corruption and a means for redressing it, his flight through the heavens offers him a glimpse of the rewards that await all righteous souls aimed at strengthening his resolve. But because he now enters a realm of divine splendor and perfection, the like of which he has never seen nor could ever easily imagine, it must be signified to him in a way his feeble human intelligence can grasp. As Beatrice observes, explaining that the spheres are not themselves the heavenly hierarchies but merely representations of them,

> It is needful to speak thus to your faculty, since only through sense perception does it apprehend that which it afterwards makes fit for the intellect. For this reason Scripture condescends to your capacity. (DSI 3:1:IV:39)

The "ad hoc command performance"[17] Dante is about to witness is only an approximation, a dramatization of the ineffable, albeit a vivid, ingenious, and often dazzling one.

Whereas hell and purgatory are visible entities in Dante's scheme, the one a pit, the other a mountain, the system of concentric crystalline spheres of the Ptolemaic universe, used quite appropriately to symbolize the heavens, is transparent and invisible. As was the case with Plato's Great Whorl, an effective way of illustrating the structure of such a complex system is to imagine it in cross section. Plato selected the metaphor of a whorl, but Dante often uses the simple image of a "wheel" (*rota*) to evoke the cross section of planets aligned along the plane of the ecliptic. His fully realized Ptolemaic symbol does nevertheless resemble Plato's Great Whorl, which is at once hemispherical (the "nest of bowls" of the whorl itself), circular (the planetary paths or "rims" of the bowls) and spherical (the implied but transcendent and thus unrepresentable other hemisphere).

In her description of the structure of paradise, Beatrice refers to the Ptolemaic metaphor as *santi giri* (II:127). The problem posed to the English translator of this term is again skirted by Sayers, whose "holy gyres" (II:66) is just as uncertain as her rendering of its infernal analog, *empi giri*, as "impious gyres." Singleton, on the other hand, who translated the latter "impious circles," commits himself more boldly in the heavenly image to "holy spheres" (II:23), an unusual translation of *giri*[18] but an appropriate one for Dante's cosmic metaphor, which is inherently equivocal[19] and which relies upon the semantic depth of language to convey the complexity of an overwhelming experience. Consequently, any attempt to appreciate Dante's brilliant vision of paradise requires the ability to tolerate ambiguity and apparent contradiction and to assimilate images not only literally but paradigmatically, as we shall see.

There are four great symbols that are of particular concern here because of their concentrically circular, spiro-helical, or vortical structures. Two of them have already been alluded to in the description of the Ptolemaic model: the "wheel" of planetary orbits and the "nest of hemispheres" (of which these orbits are the rims) that turn on the axis of the North Star. The other two are the primum mobile and the celestial rose. While it is impossible to relate these figures spatially or temporally in a fully satisfactory logical manner—the pilgrim is moving beyond time and space—certain correspondences between them can be disengaged that help to elucidate the paradoxical harmony and counterpoint of the heavenly hierarchies.

From the perspective of the planetary wheel, Dante's voyage is "horizontally oblique" along the ecliptic, relative to the celestial north-south axis, as he flies from planet to planet, and the wheel is a fitting image, for the planets' velocities increase as he moves outward, just as do velocities upon the spoke of a wheel. The case of Plato's vortical whorl is just the opposite, with velocities decreasing toward the periphery in the manner of a naturally occurring free circular vortex. Dante's system turns as a solid self-contained unit, displaying the characteristic "equiangular velocities" of a "forced vortex" (see Appendix).[20]

It may seem strange that Dante's journey toward God in the empyrean should be from center to periphery, but it makes sense in terms both of the depiction of Satan at the center of the world—"the ill Worm that pierces the world's core" (DSA 1:XXXV:288)—from which he is escaping, and of Catholic theology, which designates degrees of heavenly bliss; and the expanding planetary spheres, we learn in Canto XXVIII, are meant to depict these increasing quantities of virtue:

> The material spheres are wide or narrow according to the more or less of virtue which is diffused through all their parts. Greater goodness must needs work greater weal; and the greater body, if it has its parts equally complete, contains the greater weal. (DSI XXVIII:317)

But if Dante's celestial scheme seems to place the deity at the periphery of the universe, an ultimate circumference enclosing all, it locates him at the center too, in the image of the primum mobile, as if to illustrate literally the paradox of God as "a circle whose center is everywhere and circumference nowhere." While it is true that the pilgrim moves from planet to planet, presumably along the ecliptic, he ends up at the primum mobile, a complex of nine concentric circles telescoping out from and whirling about a single point of intense light. The point of light symbolizes God, the circles the nine choirs of angels. At the same time, the primum mobile is the ninth or outermost crystalline sphere whose movement imparts motion to all the rest and whose axis is the traditional celestial axis running north-south and centered on the polestar.[21] Polaris, then, is the common element linking both images, i.e., the divine energizing spark and the axle about which the northerly concentric hemispheres turn. Inasmuch as Dante's pilgrimage culminates in a beatific vision focused on this point, his trajectory is not only oblique along the ecliptic, but northward and vertical in the direction of the celestial axis. Add to this the fact that he begins his ascent in the southern hemisphere, suggesting a southerly course, and that he later enters Gemini, his native sign, on the (roughly) east-west axis of the ecliptic, and it becomes apparent that the exigencies of time and space are again defied, as the poet insists they must be in any attempt to portray the infinite.

Any serious stargazer knows that the constellations visible in the night sky revolve in the course of the year with the revolution of the earth around the sun. The polestar, at the zenith, however, remains rather constant, and the "Little Dipper"—what Dante calls the "horn"—extending out from Polaris, makes a counterclockwise circuit about its fixed center, like the hour hand of an annual clock in reverse. Similarly, the rotation of the earth on its axis creates the impression that stars sweep across the sky on any given night. A time exposure of the celestial zenith taken at the North Pole records, in fact, the apparent concentric streamlines of a vast celestial vortex.[22] If Dante's journey is ultimately to God at the center of the universe (the point of light of the primum mobile) and if Polaris symbolizes this primal spark, then his voyage is at once *out* through the "nest of concentric hemispheres" and *in* toward the center of the primum mobile's congruent vortex of concentric rings. Thus the movement to the periphery of the spheres may be said to be simultaneously aimed at the axial center around which they turn.

Let us consider this concurrence of images more closely, for the poet's intention is obviously to interrelate them, as disparate as they may at first seem. The "backward" forced vortex of the hemispheres, i.e., the "ad hoc command performance," is staged for Dante as part of the didactic "condescension" of perfect intelligence to his lowly human level, which is initially incapable of tolerating or assimilating the brilliance of divine illumination. His capacities must be increased gradually through education. By the time he passes into the empyrean, he is ready to see how things really *are,* which is not the way they may *seem.* In this sense, the empyreal image of the primum mobile sets things straight and shows itself to be inversely related to the system of spheres. Whereas the spheres increase in speed toward the periphery, the whirling of the angelic circuits decreases as they project outward from the divine spark,[23] manifesting the inverse radial velocities of a free circular vortex. The most dynamic choir (the Seraphim), closest to the divine center, turns the sphere of the primum mobile, the farthest from the infernal, earthly center, and so on proportionally to the outermost circuit of the divine minds, termed simply the Angels, who impart their slower motion to the innermost sphere of the moon.[24] By thus "confusing" these two major images via a vortical whirl, the poet ingeniously illustrates the complementary nature of the heavenly hierarchies, in which all apparent contradictions are resolved, and as well, the paradox of God's ubiquity.

Certainly, there can be no disputing that a vortical gyrating is the very principle of motion in the *Paradiso,* involving both a spin and/or a revolution, just as the heavenly bodies themselves rotate on their axes and in some cases revolve about each other. These gyrations are characteristic not only of the souls Dante meets along the way but of his own ascent, which, we learn in a simile of Canto V, is "as an arrow that strikes the target before the bowcord is quiet" (DSI V:55): that is, centrally aimed, instantaneously swift and direct, and spinning helically.

In the third heaven of Venus, Dante and Beatrice are caught up in the double revolution of the planet, "which is carried round by its sphere at the same time as it revolves in its own epicycle" (DSA 350–51),[25] and they gyrate with the amorous souls there in the manner described by Charles Martel: "With one circle [*giro*], with one circling [*girare*] and with one thirst we revolve [*volgiam*] with the celestial Princes" (DSI VIII:84–85). Like the whirling wind buffeting the heterosexual lovers and the circling dance of the homosexuals in hell (as well as their contrary, though complementary, circuits on the seventh cornice of purgatory), the gyrations of the celestial lovers mirror aptly their passionate ardor, but the souls in heaven whirl with the energy of pure divine love that suffuses all the spheres like electricity.

The harmonizing effect of this force is portrayed in the gyrations of the heaven of the sun. Upon entering this sphere, the poet first alludes to its spiral convolutions, remarking that it "was wheeling through the spirals [*girava per le spire*] in which it presents itself earlier every day" (DSI X:108–9). In the following cantos, Dante and Beatrice are surrounded by two concentric "melodious gyres" of theologians, which revolve in perfect unison "even as the eyes which, at the pleasure that moves them, must needs be closed and lifted in accord" (DSI XII:131). The complete harmony of their motion and song is underscored by the stories of St. Francis and St. Dominic, the first recounted lovingly by St. Thomas Aquinas, a Dominican, who emerges from the first circle of lights, the second related with obvious admiration by St. Bonaventure, a Franciscan, who speaks out of the second circle. Although often rival confraternities on earth, the generous gesture of Dominican to Franciscan and reciprocally of Franciscan to Dominican testifies to "an ideal partnership between these and other Orders" (DSA 161), as well as to the spirit of cooperation that pervades the heavenly realm.

In a final symbol of reciprocity before quitting the fourth heaven, the verbal exchange between Beatrice and Thomas provokes the following image in Dante's mind: "From the center to the rim, and so from the rim to the center, the water in a round vessel moves, according as it is struck from without or within" (DSI XIV:153). The stream of ripples, alternately expanding and contracting between center and periphery, does vividly illustrate the harmonious exchange of discourse between these exalted spirits, but the image of concentric circles also calls to mind, through paradigmatic association, the great wheel of planetary rings along the ecliptic, symbolizing, on this level, the dynamic pulsations of love that pass between the deity and his heavenly host in their vibrant, ever-increasing bliss.

When, after the dramatic encounter with his ancestor Cacciaguida, Dante ascends to the heaven of Jupiter, he comments once again on the nature of his voyage, observing, "so did I perceive that my circling round [*girare intorno*] with the heaven had increased its arc" (DSI XVIII:202–3). That he and Beatrice revolve as they ascend clearly denotes a helical pattern, but the increasing arc of their circuits reveals that their path is actually spiro-helical, as if being whirled toward the rim of a centrifugal vortex.

In the seventh heaven, the vortical imagery continues and increases. Here the whirling flames of the contemplatives descend a golden ladder, and their spokesman, the ardent spirit of Peter Damien, Dante recounts, "made a center of its middle, and spun round [*girando sé*] like a rapid millstone" (DSI XXI:238–39). Of the other souls, he observes, "I saw more flamelets from step to step descending, and whirling [*girarsi*]; and every whirl [*giro*] made them more beautiful" (DSI XXI:240–41). The whirl represents not only the energy of divine love, but also its power to purify and beautify the soul it enthralls. Finally, having heard St. Benedict rebuke the corrupt state of the monasteries, the entire host of contemplative souls seems suddenly swept away in a whirlwind: "Thus he spoke to me, then drew back to his company, and the company closed together; then like a whirlwind [*turbo*] all were gathered upward" (DSI XXII:250–51). In a manner analogous to privileged prophets of the Old Testament, the exalted status of the contemplatives is indicated by their subsumption into a whirlwind.

The gyrating persists and even intensifies in the eighth heaven, where, Dante notes,

> those glad souls made themselves spheres upon fixed poles, flaming like comets, as they whirled [*volte*]. And as wheels within the fittings of clocks revolve [*si giran*], so that to one who give heed the first seems quiet and the last to fly, so did those carols, dancing severally fast and slow, make me judge of their riches. (DSI XXIV:266–67)

Like Dervishes whirling in varying degrees of ecstasy, the glorified saints of the firmament are portrayed as the very cogs and wheels of the complex mechanism of creation—here symbolized by clockworks—and of time itself.

Nevertheless, it is in the next heaven, the outermost sphere of the primum mobile, that all time and motion originate and in which the preceding series of whirling images culminates, for Dante represents the dynamic engine of the "unmoved mover" not simply as a sphere, but as a dazzling, fiery, free, circular vortex. The structure of this symbol has already been analyzed, although its curious resemblance to modern conceptions of the atom—charged electron rings surrounding a nucleus—should also be mentioned, if for no other reason than to demonstrate the stunning prescience of artistic intuition and its, at times, uncanny correspondence with scientific fact. The *mobile primo* (XXX:107) is certainly one of Dante's most compelling images, and it fits perfectly into the overall hierarchical design of the *Comedy,* illustrating, as I have suggested, the deity at the center of all light, life, and love and showing his inverse relationship to the earth-centered system of concentric (hemi-) spheres through which the pilgrim travels to his ultimate destination, the vision of God in the empyrean.

But there remains one last mirror image, which juxtaposes the primum mobile with the celestial white rose, in what is perhaps the most ingenious and beautifully wrought symbol of the last cantica:

> A Light is thereabove which makes the Creator visible to every creature that has his peace only in beholding Him. It spreads so wide a circle that the circumference would be too large a girdle for the sun. Its whole expanse is made by a ray reflected from the summit of the Primum Mobile, which therefrom takes its life and potency; and as a hillside mirrors itself in water at its base, as if to look upon its own adornment when it is rich in grasses and in flowers, so above the light round and round about in more than a thousand tiers I saw all that of us have won return up there. (DSI XXX:341–43)

Such is the "splendor of God" and the "high triumph of the true kingdom" (DSI XXX:341) that Dante attempts to capture in his image of a verdant hill reflected in a lake, which is analogous to the implied mirror imagery of hell's pit and Mount Purgatory. Here, though, the poet seeks to depict the degrees of goodness from the celestial depths, the "yellow" of the rose into which Dante himself is led by Beatrice to behold the ascending whorls of empyreal exaltation, to the zenith where God reigns radiantly on high (although, symbolic of the pervasive nature of his light, the primal spark is also reflected in the yellow of the rose at the opposite pole).

In between, the concentric angelic circuits whirl outward and downward to the mountain's base, where the widest and outermost ring mirrors in the water the corresponding outermost row of enthroned saints in the reflected flower, whose concentric tiers telescope down to the yellow stamens at the rose's core. In an inversely analogous manner, however, the velocity of revolution decreases from the central dynamo of the primum mobile to the base, while the degree of virtuous merit in the flower's "amphitheater" increases from center to periphery. The upper hemisphere, moreover, is fiery and vortically dynamic, the lower serene and the positions fixed, although the whole is organic and growing. A paradox of this inverse double image is that it does depict a continuum of progressive nearness to God, from Dante at the nadir to Beatrice and the Blessed Virgin in the upper tiers, through the hierarchy of divine minds (from the lesser Angels to the first-rank Seraphim whirling ecstatically in the ring most immediate to divinity). And, in a final image of harmony and community, angels descend like heavenly bees to pollinate the celestial flower.[26] In such a way, Dante's lakeside hill not only embodies the structure of paradise in a recognizable natural scene, but also illustrates the medieval theology of why God created the universe: specifically, to make manifest his love by extending it to others and to increase it through mutual sharing and reflection.

The network of helical, spiro-helical, and vortical structures in Dante's *Commedia* is indisputably the most complex example examined thus far, and it is the longest single work that this study addresses. Because of its seminal importance in the history of ideas, I have felt it necessary to analyze in detail the elaboration of spiro-vortical symbolism by this poet, so that subsequent examples of the image can be more fully appreciated. My purpose is not necessarily to link the later figures to earlier ones by attempting to reconstruct a diachronic concatenation of cause and effect (even though some such links doubtless exist) but, as has been done with Dante's predecessors, to set in motion the interplay of synchronic correspondences that illuminate the symbol's paradigmatic diversity and depth.

The Turbulent Dream-Vision of Descartes's "Olympian" Experience

6

Descartes's theory of the *tourbillon* is a well-known feature of his scientific cosmology, but it is not general knowledge that the development of this theory was preceded by a strange oneiric experience involving a cryptic symbolic *tourbillon,* an experience so disturbing, so revealing to him that, by his own admission, it stood out as one of the most important of his life.

The scientist's attempt to account for planetary motion by means of the mechanics of the vortex, published in his *Principia philosophiae* in 1644,[1] has proven erroneous, but it was a focal point of scientific debate for decades, eliciting the opinions of many early modern philosophers. According to E. J. Aiton, in his meticulous reconstruction of this debate,[2] the abstract physics of the theory were more important to Descartes than technical details, and "it was the challenge presented by these details that eventually led to the downfall of the theory" (5).

This may seem ironic, since Descartes presented the first fully mechanical explanation of the celestial bodies, and his goal was to establish facts (truths) with the certainty of mathematical demonstration, in keeping with the rigorous rationalism of the scientific "method" he devised. Nevertheless, like the Pre-Socratic philosophers who had developed their own (decidedly less scientific) "cosmic whirl," he did not completely separate physics from metaphysics. In fact, the certainty of his results rested upon such metaphysical axioms as "divine immutability," so a transcendent dimension, liable to interpretation as metaphor, tends to undermine the avowed apodicticity of his aims.

Aiton notes, for example, his insistence upon a circular vortical model of the solar system, although it seems highly improbable that a thinker with the breadth of knowledge he possessed could have been totally ignorant of Kepler's findings, indicating elliptical planetary orbits. Is it possible that a cosmic model predicated upon the circle's perfect divine symmetry proved irresistible to a devout deist like Descartes, leading him to put exigencies of symbolic perfection before the disturbing inconsistencies of scientific observation? While Descartes seems clearly to have been more the theorist than the empiricist, it is nonetheless astounding to learn that "Neither Kepler nor Galileo are mentioned in any of his published works"; and Aiton adds that "if he had been more receptive to the ideas of Kepler, he might well have achieved a better formulation of his own vortex theory" (30).

Although vortical mechanics turned out to be inadequate when applied to the solar system, spiro-vorticity has since been shown to be a primary structuring principle on the galactic level, and some theorists argue that all galaxies pass through such a phase. Descartes's turbulent cosmology was thus figuratively correct, even if literally misapplied, since the vortical configuration is ubiquitous in the universe. This brings us back to the disturbing specter of the *tourbillon* that figured so prominently in the early dream experience. Whether or not that episode, recorded in a now-lost "small parchment register"[3] under the rubric *Olympica,* was, as the young philosopher strongly believed, a visionary intuition from on high into the nature of things remains uncertain, but there is no doubting its tremendous effect upon his life and thought.

The researcher who looks into the matter is soon aware of the ironic fact that the *Olympica* has created a great stir for a text that, in effect, no longer exists. Even Freud has expressed an opinion[4] about the "trois songes" ["three dreams"] described in the missing manuscript. By chance, Descartes's early biographer Baillet, who saw the text, wrote a paraphrase in French of the Latin original and thus rescued details of the event from otherwise certain oblivion. This paraphrase constitutes the core of two main editions of the work. Under the title *Olympiques,*[5] Garnier presents an intertext composed of Baillet (modernized) and relevant excerpts (also translated into modern French) of Latin notes taken by Leibnitz when he too had access to the parchment notebook. The *Olympica* in Adam and Tannery's definitive collection of the *Œuvres de Descartes*[6] is comprised of Baillet alone, although in seventeenth-century French with some minor editorial modifications. (I shall nonetheless rely principally upon the original, unaltered version in Baillet's biography.)[7] Given the confusion and uncertainty surrounding the text, from which Descartes, the original author, is almost completely absent, one might well wonder why such a "secondary source" would be placed in his collected works at all. But the truly extraordinary nature of the episode justifies preserving it in whatever form it exists.

As if the situation were not complicated enough, there exists in Descartes's account (and carried over into Baillet's) both a reconstruction of the dream events (the *rappel*) and an attempt to interpret them. In the case of Baillet's paraphrase, there are also uncertainties concerning just who, at various times, is commenting, the philosopher or his biographer.[8] The idea that interpretation can ever be separated from "facts" would probably find few adherents in modern criticism, but certainly Baillet (and Descartes) sought to distinguish between them. In fact, Baillet calls attention to the curious fact that Descartes did his interpreting asleep as well as awake:

> mais il en fit encore l'interprétation avant que le sommeil le quittât. [but he even made the interpretation of it before sleep left him]
>
> M. Descartes continuant d'interpreter son songe dans le sommeil . . . [M. Descartes, continuing to interpret his dream in his sleep]
>
> il se réveilla sans émotion: & continua les yeux ouverts, l'interprétation de son songe. [he awoke without emotion and continued, with open eyes, the interpretation of his dream] (83–84)

That one might interpret a dream while asleep certainly raises the question of just what a dream is and is not, since we have, in such a case, the apparent *mise en abyme* of dream within dream. Because, in Descartes's case, he began interpreting right after the first dream, reflecting upon its strange symbolism during "un intervalle de prés de deux heures dans des pensées diverses sur les biens & les maux de ce monde" ["an interval of almost two hours in diverse thoughts on the goods and evils of this world"] (BV 82), another question arises concerning the extent to which the two ensuing dreams were conditioned by the two hours of moral reflections separating them from the first.

One further problem that must be addressed is the order of events. Gouhier offers the insight that "l'interprétation précède tout rappel" ["interpretation precedes any (all) recall"] (34), and Alquié notes that

"M. Gouhier has rightly insisted on the importance of the fact that the interpretation of the first two dreams follows that of the third" (58n.). I must nonetheless take exception to the *tout* of Gouhier's *tout rappel.* The two hours spent mulling over the "goods" and "evils" of the world after the first dream—the first attempt at reconstruction, as Gouhier himself admits,[9] and interpretation, as the text shows[10]—do precede and undoubtedly influence the second and third dreams and subsequent interpretations, but they follow the occurrence of the first dream. This sequence of events indicates that the initial *rappel* of the first dream is the one closest to the "raw," unassimilated dream content itself, making it, as well, the dream least affected by the interpreting process, which Descartes puts into action immediately afterward. Conversely, the later "definitive" interpretation of it is the farthest removed from the dream event, since Descartes begins by explaining the third dream and works backwards (here I fully agree with Gouhier), so it ends up the dream most affected by the accumulation of allegorical significations that the Cartesian interpreter creates. Hence, as I shall attempt to show, the second and third dreams function for the perplexed thinker as, in effect, "glosses" to the first one,[11] the very strange, terrifying and ominous nature of which he is literally trying to "rationalize." Finally, my attention is drawn to the mysterious *tourbillon* of the first dream, a seminal image that overshadows and shapes subsequent events and thus merits special scrutiny.

Briefly, the experience of the "trois songes" unfolds as follows. On November 10, 1619, Descartes (then twenty-three years old), during a brief stopover in Germany, found himself *tout rempli de son enthousiasme* ["completely filled with enthusiasm"] after having discovered *les fondemens de la science admirable* ["the foundations of the admirable science"] (BV 81). Georges Poulet attributes this "quivering of the spirit" to mental exhaustion ("nervous fatigue due to extreme intellectual tension. . . . Agitation, fatigue, agitation: a rhythm of excitement and exhaustion")[12] and "the alternation of cyclothymia," a manic-depressive syndrome in which "the mind finally passes with disconcerting rapidity from an excess of joy to an excess of sadness or anguish" (63). Baillet's statement "que le feu lui prît au cerveau, & qu'il tomba dans une espéce d'enthousiasme" ["that fire seized his brain and that he fell into a kind of enthusiasm"] (81) has neuropathological overtones, echoed by Gouhier, who shows that Descartes's "feverish" condition had been developing over months: "The creative fever of March returns, but so violently that he calls it 'enthusiasm.' "

Gouhier further observes that " 'enthusiasm' is simultaneously the cause and the effect of invention" (51–52).

The chronology of the three dream episodes can be summarized quite simply: the young mathematician is, in the first dream, caught up and spun about in *une espéce de tourbillon* ["a kind of vortex (whirlwind)"] (BV 81) that throws him off course and disorients him. In the second, he is startled by a thunderclap and, opening his eyes, perceives sparks of light in his room. Finally, in the third, which Poulet describes as "the dream of deliverance, reconciliation and assuagement" (79), he opens an anthology of poets and reads the line, *Quod vitae sectabor iter?* ["What way of life shall I follow?"] (BV 83). Because Descartes's definitive reconstruction and interpretation of this "retrospective vision" (GD 33)—although engendered during the two hours of pondering after the first dream—really gets underway after the third, following the reversed chronology should yield insights and facilitate my deconstruction of the dreamer's account.

To begin with, the third *songe* seems principally to reenact the classic "conversion" experience of St. Augustine, who was likewise moved to open a book at random to discover a similarly pertinent and sententious phrase that changed his life. Petrarch, it will be remembered, duplicated the gesture on Mont Ventoux in a moment of personal crisis. Significantly, in Descartes's case, however, the volume is not a religious one but a *Corpus Poëtarum* (BV 83), which betrays the essentially "poetic" nature of the entire experience—a scientific creative fever and enthusiasm like that of poetic inspiration—and elicits the conclusion (all the more astonishing coming from a philosopher) that the insight of poetic inspiration is superior to that of philosophical reasoning.[13]

Taken together, the symbols and images of the third *songe* tend to fall into pairs of antitheses, and Descartes's interpretation of them is similarly polar and antithetical. The verse beginning "*Est & Non*" (BV 83) is, for him, "le Ouy & le Non de Pythagore" ["the yes and no of Pythagoras"] and stands for "la Vérité & la Fausseté dans les connoissances humaines, & les sciences profanes" ["truth and falsity in human knowledge and the profane sciences"] (84). Another opposition exists, symbolically, between the *Corpus Poëtarum* and an elusive "*Dictionnaire*" (82–84), the former representing "la Philosophie & la Sagesse jointes ensemble" [philosophy and wisdom joined together][14] and the latter a synthesis of "toutes les Sciences ramassées ensemble" ["all the sciences gathered together"]. Finally, the dreamer's own comprehension of the experience is described as one of doubt due to an essential equivocacy. Like Keats, at the end of "Ode to a Nightingale," Descartes, "doutant s'il révoit ou s'il méditoit" ["doubting whether he was dreaming or meditating"] (84), wonders confusedly whether it was "songe ou vision" ["dream or vision"] (83).

In many ways, the polarized organization of the dream elements may be due to the methodical structuring of them by Descartes's highly deductive mind. Even this soon in his career, his strict rational procedures are apparent, as is his tendency to conceive of things in terms of antithetical divisions. What is most striking in his recollection of the dreams, however, is the readiness with which he considers the polarities in terms of opposing moral categories, even to the point of fabricating a kind of dualistic struggle between good and evil forces in order to explain the causes behind events.

Right after the first dream, for example, during the two hours of diverse thoughts on, significantly, "the goods and evils of this world," Descartes invents the specter of a "mauvais génie" ["evil spirit"] (82) to account for the turbulent wind that accosted him. Later, after the third dream, as he constructs his reverse retrospective narrative, an opposing, complementary "Spirit of Truth" is, in a parallel manner, created to explain the positive connotations of that dream: "il fut assez hardy pour se persuader, que c'étoit l'Esprit de Vérité qui avoit voulu lui ouvrir les trésors de toutes les sciences par ce songe" ["he was so bold as to persuade himself that it was the Spirit of Truth who had sought to open to him the treasures of all the sciences by this dream"] (84). It is as if the "mauvais génie"—originally a construct—has, through the course of the night, metamorphosed into an actual dream element, spawning a counterbalancing good spirit to offset it. Baillet even seems to abet Descartes in the elaboration of his moral allegorical typing by tending to translate the Latin word *spiritus,* one of the few verbatim words we have of the lost original, as "génie" when referring to forces of evil or of unknown origin and "esprit" when associated with forces of good.[15] At any rate, these spirits become central figures in the philosopher's interpretation, even though they do not appear at all in any of the dreams themselves.[16]

Thus, the third *songe,* "qui n'avoit eu rien que de fort doux & de fort agréable, marquoit l'avenir selon luy" ["which had contained nothing that was not most sweet and agreeable, signified the future according to him"], a favorable portent "pour ce qui devoit luy arriver dans le reste de sa vie" ["for what should happen to him during the rest of his life"] (84), precisely because it develops under the auspices of the Spirit of Truth. By contrast, the "terror" and "fright" that accompanied the two preceding dreams reveal them to be "des avertissemens menaçans touchant sa vie passée, qui pouvoit n'avoir pas été aussi innocente devant Dieu que devant les hommes" ["threatening warnings concerning his past life, which could not have been as innocent before God as before men"] (84). The first of these is, of course, dominated by the "mauvais génie," who casts an ominous shadow over his past, which is as dark and guilty as his future is promising and bright.

It should not be surprising, then, that the second *songe,* wedged, as it were, between opposing forces, feelings, and values, is actually a field of confrontation and conflict, symbolized by the thunderclap and lightninglike flashes[17] of metaphorically colliding storm clouds that now beset Descartes. This dream is, on the one hand, haunted by guilt and shame for his past actions (what Baillet calls "sa syndérêse, c'est-à-dire, les remords de sa conscience touchant les péchez qu'il pouvoit avoir commis pendant le cours de sa vie jusqu'alors" ["his *syndérêse,* that is, the remorse of his conscience concerning the sins that he could have committed during the course of his life up until then"]) and which the fright carried over from the first dream denotes. But it is also the scene of saving divine intervention, since the thunder, Descartes decides, "étoit le signal de l'Esprit de vérité qui descendoit sur luy pour le posséder" ["was the signal of the Spirit of Truth which descended on him to possess him"] (85).

Hence, by the time Descartes's interpretation retraces its steps to the beginning of the dream experience, a rather rich and elaborate allegorical structure has evolved, ready-made for deciphering the all-important, cryptic, and disturbing images of the first *songe.* As I read the text, this is the puzzle that Descartes believes, consciously or not, he must solve—he did mull over it for two hours—in order to regain his calm and peace of mind.

Gouhier's condensation of the dream reduces it to its fundamental features and is thus a convenient starting point:

> The first dream is a very complicated story:
>
> 1. Some phantoms frighten him: "he was obliged to switch over to the left side in order to be able to advance to the place where he wanted to go, because he felt a great weakness on the right side, on which he could not support himself."
>
> 2. "Ashamed of walking in this manner," he wants to straighten up, but "an impetuous wind carrying him off in a kind of vortex causes him to spin three or four times on the left foot."

> 3. He drags himself along with difficulty, sees a college, enters, evidently wanting to stop; "he tried to gain the church of the college, where his first thought was to go say a prayer."
>
> 4. Noticing that he has passed a man of his acquaintance without greeting him, he wishes to retrace his steps to pay him his respects: but "he was repelled with violence by the wind which was blowing against the church."
>
> 5. At the same time, in the middle of the courtyard, someone calls him "by his name" and asks him to go over to Monsieur N. who has "something to give to him": Descartes "fancied that this was a melon which had been brought from some foreign country."
>
> 6. What prevails in him is a feeling of surprise: all those he sees around him are "straight and firm on their feet," whereas he remains "bent and wavering," even though the wind which had almost upset him "had greatly diminished."
>
> "He awoke on this imagination. . . . " Now, "he felt right then an actual pain," and "he immediately turned over on his right side; for it was on the left that he had gone to sleep and had had the dream." (35–36)

In this last paragraph, Gouhier has left out, for his own particular purposes of emphasis and analysis, a crucial clause qualifying Descartes's feeling of an "actual pain," which, Baillet recounts, "lui fit craindre que ce ne fût l'opération de quelque mauvais génie qui l'auroit voulu séduire" ["made him fear that this was the operation of some evil spirit who might have wanted to seduce him"] (82). The morally unspecified reference to "phantoms" at the beginning of the dream becomes, at the end, a specifically evil spirit, as the dreamer ventures his first interpretive remark. The metamorphosis is subtle but unmistakable: from "fantômes" (81) to "quelque mauvais génie" (morally typed but qualified by the indefinite "quelque" [82]) to, finally, "le mauvais Génie" (with definite article and capitalized letter [85]) in the definitive interpretation at the end of the whole experience. At this final stage, moreover, still another spirit is invented to explain his initial impulse toward the church: "L'Esprit de Dieu qui luy avoit fait faire les prémiéres démarches vers cette Eglise" ["the Spirit of God who had made him take the first steps toward that church"] (85). In such a way, the first *songe* is transformed by Descartes into a morally charged moment of Faustian wavering between competing influences of a "good" and a "bad" spirit, the former impelling him toward the church, the latter rebuffing his efforts to redress his incivility toward the man he has snubbed by forcing him against his will to continue in the direction of the church.[18]

This dichotomy runs through the entire dream sequence, depicting what Poulet calls "a symbolic image of life divided in two" (67). In fact, the episode is, to a large extent, structured in terms of a polarity of left and right,[19] which certainly conduced readily to the dreamer's subsequent moralistic reading, given the traditionally negative ("sinister") associations with left and positive, veracious connotations of right. These significations are borne out in the dream elements. Immediately at the outset, "il étoit obligé de se renverser sur le côté gauche pour pouvoir avancer au lieu où il vouloit aller, parce qu'il sentoit une grande foiblesse au côté droit dont il ne pouvoit se soutenir" ["he was obliged to switch over to the left side in order to be able to advance to the place where he wanted to go, because he felt a great weakness on the right side, on which he could not support himself"] (BV 81). Weakness on the right, a countermove to the left, and suddenly, he is caught up in a whirlwind and finds himself executing three or four pirouettes "on the left foot." At the end, having awakened, "il se retourna sur le côté droit, car c'étoit sur le gauche qu'il s'étoit endormi, & qu'il avoit eu le songe" ["he turned over on his right side; for it was on the left that he had gone to sleep and had had the dream"] (82). In other words, he rolls over on his right side because falling asleep on his left side must have induced this "bad dream," in which his right side was weakened and overpowered by "sinister" forces. The moral implications of the *songe* are thus cogently suggested by right- and left-handed symbolism, even without Descartes's interpretation of it in terms of good and evil.

In a parallel manner, an opposition between "self" and "other" pits Descartes, who describes himself as "courbé & chancelant" ["bent and wavering"], against certain unnamed others, who are "droits & fermes sur leurs pieds" ["straight and firm on their feet"] (82). Here, the steady, upright stance of the strangers casts the dreamer's stooped, staggering demeanor in a negative light, and the double meaning of "droit" links "straight" posture with the positive connotations of "right" in the images just analyzed, both of which emphasize the philosopher's own curved (crooked) and leftward course. The encounters with the two shadowy men serve further to censure Descartes's behavior. The first "man of his acquaintance" whom he passes "without greeting" (81) not only points up his lack of civility but also tends to render his receipt of the second

Monsieur N.'s gift of a melon[20] all the more undeserved. Thus Descartes's selfish lack of common courtesy is compared unfavorably to another's friendly generosity.

There are still other dualisms in the passage that could be discussed—for example, the twin destinations of "college" and "church" (a conflict between his scholarly aspirations and religious duties?), opposing feelings of "enthusiasm" and "guilt" and so on—all of which establish beyond any doubt the polarized, antithetical structuring of the dream events. What is most significant, however, is that their confrontation produces, like cross currents clashing in a flow, that vertiginous synthesizer, the vortex.

Of all the figures in all three dreams, the *tourbillon* is, without a doubt, the most astonishing, dynamic, and disruptive one. Its aggressive, menacing role in the first dream certainly does facilitate confusing it with some sort of being and, at the same time, predisposes it to speculation about its symbolic meaning. Descartes, in fact, wastes no time personifying and explaining it, engendering thereby the cast of hostile and beneficent spirits he assembles to give sense to his troubling nightmare. But in order to determine whether or not the dream elements themselves actually support his interpretation, we must first reckon with its highly religious and moralistic tone:

> Le vent qui le poussoit vers l'Eglise du collège, lorsqu'il avoit mal au côté droit, n'étoit autre chose que le mauvais Génie qui tâchoit de le jetter par force dans un lieu, où son dessein étoit d'aller volontairement. C'est pourquoy Dieu ne permit pas qu'il avançât plus loin, & qu'il se laissât emporter même en un lieu saint par un Esprit qu'il n'avoit pas envoyé: quoy qu'il fût trés- persuadé que ç'eût été l'Esprit de Dieu qui luy avoit fait faire les prémiéres démarches vers cette Eglise.

> The wind that pushed him toward the church of the college, when he was having difficulty on the right side, was nothing other than the Evil Spirit who tried to throw him by force into a place where it was his intention to go voluntarily. That is why God did not permit him to advance further, and let him be carried off, even into a holy place, by a Spirit he had not sent: although he was very persuaded that it had been the Spirit of God who had made him take the first steps toward that church. (BV 85)

At first glance, Descartes's account might seem to describe, allegorically, a theology of evil in which God intends to accomplish a good end (movement toward the church, which was divinely inspired) through evil means (the agency of the "impetuous wind"). Such a view has Protestant overtones, as Luther's analysis of the "hardening of Pharaoh" in *The Bondage of the Will* exemplifies,[21] and would be dangerously heretical for a Catholic like Descartes to espouse, especially in Europe of the Counter-Reformation. Aware of this ramification, perhaps, Descartes adds a disclaimer: "C'est pourquoy Dieu ne permit pas qu'il avançât plus loin, & qu'il se laissât emporter même en un lieu saint par un Esprit qu'il n'avoit pas envoyé." No such intervention occurs, however, in the original *rappel.* Actually, its belated inclusion only complicates the situation, since God first motivates the dreamer's course toward the church only to interdict it when the evil force does likewise. And why, in any case, is he being forced to do what he intended all along to accomplish voluntarily? In the end, Descartes's interpretation may well create more problems than it resolves and is thus not the most satisfying and certainly not the only plausible one. It would therefore seem profitable to entertain other possible significations.

One might admit his contention that the dream has cosmic and religious significance without subscribing to his literal reading of the details. The very idea of a whirlwind personifying a powerful, supernatural being does have, on the face of it, biblical overtones.[22] The image of Yahweh visiting destruction upon his enemies or venting his wrath at the faithlessness of his people "from out of the whirlwind" is one Descartes would doubtless be familiar with, and his *syndérêse* indicates a sense of guilt which would account for the appearance of a turbulent, threatening symbol of God in the *songe.* Moreover, by associating the whirlwind with the "Esprit de Dieu" and not with some mysterious evil spirit,[23] the problem of why it should impel him toward the church, instead of away from it, would be cleared up. Also, Descartes's alleged desire voluntarily to attain the church is contradicted by his wavering—"chancelant"—conduct and the ease with which he is distracted from his goal by passers-by, eliciting a forceful divine push—grace?—to strengthen his resolve, even though ultimately his weakness does prevail. Poulet's suggestion of a kind of prideful self-preoccupation, at the expense of his religious duties would explain why he was disturbed by a feeling of *syndérêse* in the first place:

> But everything indicates that in the unprecedented ardor of his intellectual "faith," and especially in those last hours of fever, discovery and enthusiasm, Descartes was so thoroughly absorbed by the search for, the conquest, the presence of truth, that the presence of God was, so to speak, blurred and effaced in his soul. Let us not even say that he forgot God. Let us note simply that God is never named in any of the accounts that refer to the moment of discovery. Now on the part of a serious Christian like Descartes, that is a significant omission. (65)

Pursuing the biblical associations further, we have, in addition to the vortical image itself, the idea of "being caught up in a whirlwind." Both Elijah and Ezekiel, as we have seen, were swept up in the divine vortex, signifying literally a subsumption into godhead and testifying to their privileged status as prophets mediating hermeneutically the divine word. In Descartes's case, the discovery of an "admirable science," Poulet asserts, "was no longer the solution to a certain problem, nor even the discovery of the general principles of a science, but the discovery of the unity of all the sciences" (61), indicating an almost visionary glimpse of truth on the part of the young scientist, not unlike that of the prophets, which his own subsumption into the *tourbillon* would seem appropriately to symbolize. The loftiness of the entire enterprise, in fact, is emphasized by Descartes himself in his insistence that the three dreams resulting from his enthusiastic discovery could only have come "d'enhaut" ["from on high"] (BV 81) and in his choice of the cryptic title, *Olympica,*[24] to describe the extraordinary events.

There remains still one more inference to draw from this remarkable text, insofar as it is the record not only of a "conversion" (i.e., transformation from a state of doubt to one of reaffirmed faith and conviction), but also of the dizzying raptures of intellectual insight and scientific discovery, comparable, as even Descartes himself admits, to the intuitions of poetic-aesthetic inspiration. At the heart of the experience: a turbulent, disorienting, intoxicating enthusiasm, events unfolding "with a rapidity so vertiginous that one has trouble embracing the full expanse of it in a single glance (PE 87); the sensation, in other words, of a kind of "creative vertigo." But the mountain image, evinced by the title, also suggests "an ascent" (KD 175) and the "acrophobic vertigo" of the towering vantage from the Olympian summit, at once frightening and exhilarating. In this way, Descartes's dream seems to prefigure the vertiginous "philosophy of the heights" of, for example, Nietzsche's Zarathustra (and later Nietzsche himself, thundering from that quaint lookout, Sils-Maria, perched so precariously on sheer Alpine cliffs), as well as more modern existentialist thinkers like Heidegger and, to a degree, Sartre.[25]

Even when viewed from the distant perspective of more than 350 years after the fact, the episode of the "trois songes" loses very little of its strangeness and wonder, notwithstanding modern demystifications of the dream process by psychology and, in particular, psychoanalysis.[26] The objectivity that such distance affords does, nonetheless, put Descartes's interpretation of the event in relief, and we may tend perhaps too readily, from our postmodern standpoint, to dismiss it as naïve and obsessively logocentric. There is no doubt that the philosopher sought to impose a rational order on even the most patently irrational elements of the dream content,[27] and I think I have demonstrated, in my deconstruction of the text, the extent to which his reading was the product of a certain "set" that led him to invent some elements and distort others in his attempt to discover in the images an allegory with a clear moral message sent to him as a sign of divine concern for him and for the course his life was taking. To regard a dream as an omen would certainly have been a very normal reaction in his day, and all the more so for an impressionable genius in search of a mission.

I have offered some variant interpretations of the dream elements not with the intention of establishing them as the "right" ones, opposed to those of the philosopher which are, accordingly, "wrong." I have been interested, rather, in affirming the rich polyvalence of the symbols (and particularly the *tourbillon*) in response to Descartes's excessively rigid, univocal, and authoritative attitude toward the interpretation of them. Actually, I am inclined to think that dream images are, for the most part, nonreferential, empty signifiers and that there is more to learn about Descartes in the way he goes about explaining and justifying them than in the specific content of the dreams themselves.

What I find particularly striking in this extraordinary text is the image of a brilliant young thinker possessed by self-doubt, fear, and guilt, heightened to a feverish pitch, culminating in an oneiric experience of intense spiritual crisis and catharsis. At the center of it, a profound conflict between science and religion, knowledge and faith, vanity and deference to divinity. Earlier, I alluded to the dreamer's Faustian wavering, since in essence Descartes's dilemma is not unlike that of Doctor Faustus. The quest for knowledge was insatiable for Marlowe's ambitious overreacher, who paid for it with the loss of his soul. There are forbidden objects of knowledge, according to this legend, echoing the drama of original sin committed near, significantly, the Tree of Knowledge.[28] Descartes's enthusiastic scientific inquisitiveness risked encroaching upon matters exclusively divine, or so the young thinker may have feared guiltily. With the papal condemnation of Copernican ideas and Galileo's espousal of them only three years past—the Inquisition's persecution was yet to come—it should not be surprising that Descartes would entertain grave doubts about transgressing the limits of scientific inquiry. In this respect, the trauma of November 1619, "cette affaire, qu'il jugeoit la plus importante de sa vie" ["this affair, which he judged the most important of his life"] (BV 85), marks a pivotal "turning point"[29] in his life, as the pirouettes executed in the first *songe* literally portray. And the text that records it is a veritable matrix of issues and images that permeate the rest of his career. The discovery of "la science admirable," for example, may well have comprised the crux of the "method" he would advocate almost two decades later in the *Discours* (although the critics are evenly divided on this point). Poulet feels that the "whirling movement" of the first *songe* "already is like a rapid evocation of what Cartesian physics will be" (68), and the whirlwind's "mauvais génie" does, in any case, prefigure "le malin génie" of the *Méditations.* Various dualistic divisions in the dreams, moreover, anticipate such important antitheses as body and soul, doubt and certainty, dream and reality—even the coordinates of a Cartesian graph are *co*-ordinates—to mention only some of the major seminal themes.

Apart from its importance in his life, Descartes's *Olympica,* few would deny, records a bizarre, riveting, illuminating experience that continues to bewilder and fascinate posterity as much as it did the great philosopher himself. For a scant twelve pages penned on parchment in a long-lost notebook, it is indeed a significant and compelling, if equally ironic, absence.

"... That Every Thing Has Its / Own Vortex ..." Dialectics of Vortical Symbolism in Blake

7

Whereas the turbulent vision of the Olympian experience stands as an anomalous poetic episode in Descartes's otherwise highly ratiocinative career as a philosopher-scientist, the role of visionary poet and prophet is central to the work of William Blake, whose repudiation of the "Satanic rationalism" of science is well known and whose use of the vortex symbol represents a pithy, if at times cryptic, synthesis of many of the vortical images examined heretofore. But although Blake's adaptation of the symbol relies upon traditional religious connotations, from the Bible through Dante and Milton, as well as upon the scientific vortex theories of the Cartesians, he frequently imbues these traditional meanings with elements from his own recondite iconography.

The abstract and even hermetic nature of the immense allegory that the English poet-prophet created—he found established mythologies unsuitable for conveying his vision—has been faulted by those who deem it too private and personal a system for any but the select few willing to master its myriad significations. This complexity certainly does render any attempt to comment upon his symbolism a forbidding task, especially when Blake is not the unique focus of inquiry, as is the case in the present study. Northrop Frye argues, however, that Blake intended his works, even the difficult prophetic epics, for the "Public" (and particularly for "enthusiasts of poetry")[1] and that there is nothing inherently "mystical" about his verse, although the poet does deliberately avoid overly explicit statements in favor of an appeal, through allegory, to the "divine human faculty" (the imagination). It should not be impossible, then, to capture a sense of Blake's vortex without becoming enmeshed in the "tangle of encyclopedias, concordances, indexes, charts, and diagrams" (FF 9) that now surround his art, if attention is focused upon only the most salient and relevant texts. But here the concept of "text" is a peculiarly Blakean mélange of poetry and graphic design (sometimes one or the other or the two together), since his prowess as a wordsmith is equalled—opinions vary on this point—by his talents as an illustrator and engraver. And the vortex[2] is a ubiquitous image in both dimensions of his art.

One of the more comprehensive attempts to survey the spiro-vortical structures, especially in Blake's drawings, is W.J.T. Mitchell's, under the heading, "Metamorphoses of the Vortex." Mitchell declares the symbol to be "one of Blake's most pervasive and protean schemata"[3] and detects essentially three main connotative types:

> [1] the imagery of dissolution, annihilation, and disorientation . . .
> [2] images of dialectical interaction or conflict . . .
> [3] designs which suggest epiphany or visionary breakthrough into a new level of consciousness. (69)[4]

Mitchell's categories are themselves a striking model of dialectic, types 1 and 3 indicating, respectively, negative and positive polar antitheses, and type 2 a conflictual synthesis. When specific examples of the vortex symbol are scrutinized, however, they often seem to embody various combinations of the categories, because, according to Mitchell, "Blake thinks of them as metaphorically connected."

One of the most persistent manifestations of the symbol is what might be termed "the vortex of Urizenic, Satanic rationalism." A celebrated episode from Night VI of *The Four Zoas*[5] is a particularly cogent case in point. Urizen, "the Prince of Light" (BFZ 12:30), is obviously Blake's personification of Lucifer (Satan), and Night VI recounts his turbulent "fall":

> . . . then turning round, he threw
> Himself into the dismal void. falling he fell & fell
> Whirling in unresistible revolutions down & down
> In the horrid bottomless vacuity falling falling falling . . .
> (71:20–23)

Ultimately, in the course of his chaotic plunging and wandering through the void, Urizen attempts to rebuild his shattered world, "Creating many a Vortex fixing many a Science in the deep" (72:13). As Raine,[6] Nurmi,[7] and Ault[8] have pointed out, Urizen's new world is, in essence, the cosmos of tangential Cartesian vortices (the phenomenal, material world of human existence), and Blake is drawing upon the current hypotheses of his day, effecting, according to Ault, a poetic synthesis of the ostensibly mutually exclusive physical theories of Descartes and Newton:

> In Newton's void, vortexes cannot be sustained. In Descartes' system of vortices, there is no void space because the vortices operate conjunctively and are interconnected; Blake has transformed this image into a cosmology in which vortices can operate independently, but, once set into operation, assume all the characteristics of the Cartesian vortices. (149)

The image, near the end of the episode, of "a dire Web" that "Followed behind him as the Web of a Spider dusky & cold / Shivering across from Vortex to Vortex" (73:31–33) has a distinct Cartesian ring to it, recalling, as Ault claims, the web-like "illustrations of Descartes' vortices which appeared in eighteenth-century editions of Descartes' *Principia*" (149)[9] and evoking as well the typically spiral structure of a spider's web.

Inasmuch as Urizen represents for Blake the Satanic creator of the mechanistic principles of our cosmos, he also epitomizes the operations of reason and science, which, suggests the author, entrap and paralyze the human mind in the fallen state, blinding it to the divine eternal truths that only the imagination can grasp. The Urizenic vortex is thus a negative symbol, and "Blake implies that the Cartesian-Newtonian mechanistic universe is the true metaphysical hell" (RT II:82).

Still, "no matter how evil or mistaken [Blake] may think some character, he always envisions a divine comedy of forgiveness, reconciliation, and transformation at the end of the vortex" (MB 70). Take, for example, the spiral scroll on which the stooped figure of Newton inscribes geometrical forms with a compass, in *Newton* (MB pl. 22). On the one hand, it stands for the constricting spiral of rationalism and portrays Newton clearly as a Satanic type comparable to the figure of Urizen on the frontispiece of *Europe* (MB pl. 24), who, compasses extended, stoops from on high, like the Almighty in *Paradise Lost,* "to circumscribe / This universe, and all created things" (MP VII:226–27).[10] When, on the other hand, Newton "has persisted in his heroic folly long enough . . . he will become wise, all his abstract reductions rolling up in a prophetic scroll . . . coiling into a vortex which leads to a new level of perception" (MB 73).

Another ambivalent example involves an opposition between Jehovah (Yahweh) and Jesus. The poet's iconography generally depicts Jehovah, the jealous, vindictive God of the Old Testament, as a demonic-Urizenic figure. To Blake, the rule of vengeance, punishment, and retribution is inimical to the very concept of divinity, and it is Jesus, the revolutionary messenger of love and forgiveness, not the angry Jehovah, who is the true God. This distinction between the Old and New Testament deities is compellingly portrayed in the illustrations to the Book of Job. In plate 11, entitled "Satan with cloven hoof,"[11] the Urizenic false God hovers menacingly above Job, pointing with one hand to the stone tablets of the Ten Commandments behind him and with the other toward hell below, over which Job lies precariously. The cloven left foot reveals the true Satanic identity of this false deity, as does the spiro-helical "serpent of Materialism" in which he is entwined. Further emphasizing the "sinister" implications of the tableau, Job rests the left side of his head on a pillow formed by the extremity of his mattress rolled up in a witherwise spiral. Here, Blake seems to draw upon traditional associations of left-handedness and serpent imagery to connote evil, error, and temptation. In both cases, it is the spiral or the spiral helix that furnishes the underlying structural form.

In marked contrast, plate 13, "The Lord Answering Job from the Whirlwind" (DJ 36–37), depicts a benevolent figure emerging from a whirlwind of angels, cruciform arms outstretched, revealing him to be, as Damon points out, Jesus, the true God. But he displays the flowing beard of Jehovah, a vestige of the cruel, evil persona of plate 11 who has metamorphosed into the loving, beneficent image of plate 13. This is both an original and accurate illustration of the story's contradictory portrayal of the deity, who first allies himself with Satan (hence, the nightmarish devil-god of plate 11) to test Job mercilessly, only to heap lavish rewards upon him in the end for his fidelity (plate 13's kind father-figure with open arms ready to embrace). Symbolically, the spiro-vortical image displays a corresponding dialectical reversibility, representing, in one case, the destructive Satanic paradigm of the coiled serpent and, in the other, the vision of God in "the Whirlwind of the mystical ecstasy."

As is the case with the Job drawings, Blake often grafts elements from his own system on to the stories he has chosen to illustrate. Plate 11, just discussed, is a prime example of this "corrective" technique. Another one is plate 3, "The Mission of Virgil," in Blake's illustrations to Dante's *Comedy.*[12] Roe convincingly relates the depiction of Yahweh in this design to that of Job 11:

> That we have here the gate of the Fallen World is indicated by the figure of Jehovah with arms outstretched which dominates the top of the page. Lest we should miss his significance, Blake has written lightly in pencil above him, "The Angry God of this World." In his bearded form, we can recognize the demonic Urizen as often portrayed by Blake. Although the details are not distinct here, a comparison with the drawing from the *Paradise Regained* series, "Christ's Troubled Dream," will indicate that Blake intended to show serpents—symbols of materialism—as extending across the shoulders of the figure and hanging down from his hands, probably accompanied by forks of lightning. It will be noted that the right foot of the figure is human, but that the left is the cloven foot of an animal. The figure is much akin to that of Jehovah in the eleventh illustration to the Book of Job. (51)

Roe's claim that Blake "intended to show serpents" may overstate the case, since the poet's symbolism in one drawing cannot be adduced automatically to prove his intentions in another. Certainly, the serpentine spiral helix is suggested, but more ostensibly, as Grant claims, Yahweh "wields conical vortices"[13] in his hands, reflecting his readiness in the Old Testament to unleash a whirlwind against his enemies, a theme echoed by Dante's own insistence upon retributive justice in the divine scheme.

On the other hand, the outstretched arms recall the cruciform Jesus of Job 13, which seems incongruous in this negative Satanic image. In effect, there are many equivocal elements in this drawing, not the least of which is the male-female priest-prince-harlot (symbolizing the "prostitution" of civil and ecclesiastical authority) who kneels backward before Yahweh, his/her arms likewise extended in a gesture of submission. Perhaps Blake is attempting to dramatize by means of this dialectical ambivalence the church's, the state's, and Dante's mistaken belief in the doctrine of the trinity, which equates the Satanic Jehovah with the divine Jesus, whose outstretched arms in Job 13 are straight, the hands free and relaxed, while here, as if in parody, the arms are menacingly contorted, and the hands (more like claws) grasp deadly serpent-vortices.

One final arresting, even startling, allusion to "the angry (false) God of this world" is plate 46 of *Jerusalem*, "The Plow of Jehovah" (RT II:pl. 181). Once again, left-handedness (for example, the left-handed direction of the plow, which seems rather to be a cart upon which a forlorn man and woman ride) combines with serpent images to create a frightful, sinister effect. The plow-cart is actually fashioned from the serpent bodies, one of which coils up witherwise to form the cart's wheel. Two grotesque unicorns, with men's heads, Urizenic flowing beards and animals' bodies and hooves, pull the vehicle forward, their serpentine horns twisting helically in opposite directions, a hand attached mysteriously at the end of each one. In the entire Blakean repertory, there is no more harrowing a portrayal of the fallen Satanic state, as symbolized variously by the spiral, the spiral helix, and the vortex.

Another expression of the vortex symbol, as pervasive as the Urizenic type, is Mitchell's third category, connoting a transcendent experience or "visionary breakthrough into a new level of consciousness" (previously apparent in Job 13). The most celebrated evocation of this "epiphanal" vortex is found in *Milton*:

> The nature of infinity is this: That every thing has its
> Own Vortex; and when once a traveller thro Eternity.
> Has passd that Vortex, he percieves it roll backward behind
> His path, into a globe itself infolding; like a sun:
> Or like a moon, or like a universe of starry majesty,
> While he keeps onwards in his wondrous journey on the earth
> Or like a human form, a friend with whom he livd benevolent.
> As the eye of man views both the east & west encompassing
> Its vortex; and the north & south, with all their starry host;
> Also the rising sun & setting moon he views surrounding
> His corn-fields and his valleys of five hundred acres square.
> Thus is the earth one infinite plane, and not as apparent
> To the weak traveller confin'd beneath the moony shade.
> Thus is the heaven a vortex passd already, and the earth
> A vortex not yet pass'd by the traveller thro' Eternity.
> (15:21–35)[14]

Ault refers to this passage as "one of the most compelling and complex in all of literature" (154), and his understanding of it emphasizes the influence of "important eighteenth-century Newtonian and Cartesian optical theories concerning the horizon and the conical angle of the eye" (155). The Cartesian theory that "light consists of a series of fluid vortices by means of which perceptual objects are conducted to the eye" (158), coupled with the popular account of Newtonian optics as the vertex of a light-emitting cone, would help explain the ostensibly obscure perspectivism of lines 28–33, especially if, as Ault argues, Blake has "ironically reversed" the model so that "the vertex of the cone is in the eye itself and the cone extends outward" (159–60).

But Blake's description here is only quasi-scientific, and it would seem, on the figurative level, that every object (an immanent "minute particular" in his vocabulary) is like a point at the vertex of a cone—"every thing has its / Own Vortex"—and gaining knowledge of that particular is like passing from the distant periphery of the vortex to its essential center, drawn through and beyond by the "centripetal force" of divine poetic imagination. In this sense, Blake is describing, in lines 21–24, just what it would be like to pass through such a vortex. Issuing from the center into the unknown beyond, one would look back and perceive a circular-spherical form with the concentrically circular vortical streamlines "infolding" toward the center. The divine unity and simplicity, i.e., the "holy" form that is the essence of the particular, is symbolized by the circular-spherical "globe" and "sun." Moreover, the poet Milton (as well as Blake himself) is that "traveller thro Eternity" who is capable of "getting behind" the appearance of things by means of the poetic imagination, not the rationalistic Urizen, who, ensnared in the web of material vortices of his own making, "cannot pass through a vortex so that he may attain a perspective on his perspective . . . [and] see the three-dimensional cone from beyond one end, thus making it appear to be a disc, the sun."[15]

Hence, while "'passing the vortex' of a thing, whether a stone, a flower, or a person involves an entry into the interior life of that object, a recognition of its inherent 'genius'" (MB 71), i.e., "see[ing] an object's own point of view,"[16] the perceiver-traveller, himself a minute particular engulfed in his own vortex, must at the same time shed and transcend that vortex. In this sense, a vortex symbolizes "one's perspective from a particular space-time complex" (GT 79) or "the gateway into a new level of perception . . . the opening into infinity, a Jacob's Ladder spiraling into the heaven of heavens" (MB 73) or, in Blake's "psychological allegory," "a configuration of attitudes and events corresponding to what people mean when they say they are 'going through' something. In a broader sense, a vortex is a way of looking at things, an orientation, a pathway through chaos."[17]

The illustrations are particularly replete with images of this "transcendent vortex." Most often, the shift in perspective denotes the passage from one state of the Four Zoas to another in the context of Blake's "Circle of Destiny." This change may have positive or negative connotations, depending on the outcome. Urizen's whirling plunge in Night VI is a malevolent fall into "Ulro-void," the lowest state in the poet's visionary cosmology. Similarly, Blake's depiction of "The Expulsion" (MB pl. 11), in his illustrations to *Paradise Lost,* displays an overtly emblematic vortex swirl above and behind the departing figures of Adam and Eve, as they are ushered out of Eden by an angel. This menacing vortex symbolizes their fall from a happier state into the Urizenic realm of materialism and rationality, as the serpent slithering out through the angel's feet unmistakenly denotes. Even Milton's own vortical descent from eternal repose in Beulah back down into the mortal, annihilating world of time and space is a regression, as the dark, stormy vortex clouds on the title page of *Milton* powerfully suggest (MB pl. 18).

Nonetheless, Milton's mission, like Blake's, is "To Justify the Ways of God to Man,"[18] and his passage through the threatening vortex is therefore a heroic act. As the incarnation of divine poetic imagination, he is capable of getting at the true, essential vortical core of things. The Miltonic vortex is thus a dialectically ambivalent symbol, effecting, on the positive side, a spiro-helical contact between heaven and hell, eternity and death, imagination and reason, truth and illusion, but at the expense of a negative vortical reversion.

A less ambiguous depiction of transcendence is "Jacob's Ladder."[19] Although the form of this image is unspecified in the Bible, in Blake's rendering of it, angels and mortals harmoniously ascend (counterclockwise) or descend (clockwise) a spiro-helical staircase. The connotation of a dynamic spiro-vortical link between states or stages could not be clearer. Similarly, in the richly allegorical painting entitled "The Epitome of James Hervey's Meditations among the Tombs" (FB pl. 89), a serpentine, swirling stair connects divinity (at the vertex of both a gothic arch and the stair's spiro-vortical cone) with the mortal realm below. Finally, Blake's illustration of "Ezekiel's Vision" (FB pl. 57), true to the biblical account examined in a previous chapter, associates the vortex directly with the prophetic vision and is of particular importance to the poet, in whose vast allegory the quaternary functions as *an* if not *the* essential structure.[20]

This last painting may help to explain one of the most intriguing and cryptic vortices in Blake, which figures centrally in his mysterious depiction of "Beatrice Addressing Dante from the Car" (plate 88 in Roe's collection of the Dante drawings). Two other notable vortical images in this set, "The Circle of the Lustful: Francesca da Rimini" (plate 10) and "Jacopo Rusticucci and His Comrades" (plate 29), are more or less literal renditions of, respectively, the punishing wind that buffets carnal sinners about in *Inferno* V and the disorienting whirling dance of the three homosexuals in *Inferno* XVI (both discussed earlier in the chapter on Dante). As such, they do not shed much light on the ostensibly anomalous vortex that serves as Beatrice's car in the pageant atop Mount Purgatory.

The replacement of the wheel by a spiro-vortical form recalls Blake's approach in "The Plow of Jehovah," where the coiled serpent of materialism serves a similar symbolic function, as we have seen. Here, it is a vortex, not serpentine spires, that seems to impute a negative sense to the car upon which Beatrice rides. Do we have here the Urizenic-Satanic vortex that, like a spider's web, entraps and paralyzes those who come within its influence? The three female figures spinning about in the streamlines do seem thus ensnared. Add to this the fact that the car, in Dante's allegory, represents "the triumph-car of the Church,"[21] and it seems quite probable that Blake has sought to "correct" Dante's scheme by unmasking the true Satanic nature of the Church and of ecclesiastical law, signified by the open book that issues in a cloud of smoke from the vortex.

Roe emphasizes the feminine elements in the tableau—there are seven in all, including Beatrice, the three inside the vortex and three more standing (dancing) alongside—and he detects in the scene, what with Dante bowing sheepishly at the extreme right, "an instance of the Poetic Genius humbling itself to the Female Will" (167). Anne Mellor concurs with Roe's reading:

> This Beatrice is Blake's female will or Rahab, the fallen state of Vala. Her spiked golden crown (substituted for Dante's olive wreath), her book of faith or law, and the self-enclosed vortex that serves as the wheel of her car all iconographically establish her negative identity as the female will incarnate.[22]

Beatrice's sharp-tongued, haughty rebuke of Dante at this climactic first meeting in the *Comedy* would seem to denote for Blake the conflictual relationship between man (Dante) and his feminine emanation (Beatrice) in the fallen state, where the original unity has separated into an antagonistic dualism. Moreover, Beatrice's "clinging transparent garment" (RB 168) does imbue her with a sensuality that suggests the feminine seductive allure of the temptress, who, according to scripture, precipitated the fall in the Garden of Eden, an analogy which Dante clearly implies by his setting of this scene in the garden of the earthly paradise.

But Dante's intentions have obviously been subverted, if we accept the preceding "misogynist" reading of Blake. If it was through a man's love for a woman that humanity fell originally, it is through a higher, spiritualized version of that love, according to Dante, that humanity is saved. And this idea is not altogether inconsistent with Blake's view of "the perfect marriage" between a man and his feminine emanation in Beulah, the harmonious state of repose between the Eternity of Eden and the fallen material world. "Seen from above, Beulah is a descent into division and dreams, but seen from below, it is an ascent into an ideal, which opens the way into Eternity" (DB 367). As Roe notes, "Blake continually associates Purgatory with Beulah" (167). Thus, it is possible that the vortex, depicted here at the summit of Purgatory-Beulah, signifies dialectically both the negative Urizenic plunge into chaos and the transcendent spiro-helical path into Eternity.

Viewed in this light, Blake has preserved Dante's own perspective, even if he has grafted onto it elements of his own. Accordingly, the vortex over which Beatrice regally presides prefigures the spiro-helical journey through the heavenly spheres that she and Dante are about to undertake in the *Paradiso*. This is a figurative, visionary voyage, but the vortex is precisely the symbol of the visionary prophetic experience. The relevance of the four beasts, both in Dante's text and Blake's illustration, is therefore apparent. Echoing the case of Ezekiel, the rectilinear disposition of the quaternary combines symbolically with the curvilinear whorls, here interspersed with eyes as in the Bible, to intimate a marriage of the earthly and heavenly domains by means of the visionary transcendence of poetic imagination.

Such a positive interpretation of this important conjunction of Dante's and Blake's schemes must not, in any case, be altogether discounted, especially since Blake has rendered the scene with what would seem to be intentional ambivalence. There really is nothing overtly "sinister" about this design (the gryphon even pulls the cart "to the right"). The grotesque, terrifying impressions conveyed by "The Plow of Jehovah" or Job 11 are quite foreign to the atmosphere of dignity and serenity encountered here, which recalls, rather, the tranquil benevolence of Job 13. Beatrice's attitude is one of openness and stately repose, not at all threatening like her appearance in Dante's poem, and Dante, otherwise fully erect, bows his head reverently in prayer, not with remorseful guilt. The accompanying feminine figures evince a correspondent sense of equanimity, even those floating in the whirl, and the three dancers display a Botticelli-like elegance and grace. The riveting vortex, in fact, resembles more a whorled shell and, like that, has sensual overtones, suggesting not a direful abyss, but an inviting orifice-refuge. Such a reading would be consistent, too, with Blake's notion of the harmonious, complementary and edifying nature of feminine-masculine sexuality on the level of moonlit Beulah repose.

"Beatrice Addressing Dante from the Car" thus brilliantly epitomizes the equivocal, dialectical function of the vortex symbol in Blake's art, reflecting as well the persistent ambivalence of the image throughout the examples studied in the second part of this study. In the cases of Dante, Descartes, and Blake, these antithetical meanings are apt to be associated with Christian moral categories of good and evil, although Blake's maverick theology, while essentially Christian in its focus on Christ, radically challenges the traditional ethics mirrored in Dante and Descartes and depicts the Church itself, by means of negative serpentine and vortical whorls, as steeped in the very evil and materialism it purports to condemn.

In another reversal, Blake seems to echo the early Christian, Neoplatonist conception of the spiral (spiral helix) as denoting, psychologically, the reasoning process,[23] but he interprets it, as such, to be a thoroughly negative symbol. When, however, it stands for the convergence of earthly rectilinearity and heavenly curvilinearity, as it also does in Neoplatonist thought,[24] Blake charges it with the positive connotations of the prophetic vision. Here, both Dante and Descartes seem to agree. The spiro-vortical image figures prominently in their own transcendent experiences, as we have seen, and it also has for them a double significance, representing a plunge as well as an ascent for Dante, a burst of creative enthusiasm and a haunting specter of guilt for Descartes.

The symbolism of left- and right-handedness, moreover, is integral to the vortical imagery of all three authors. Indications of direction in a circular or spiro-helical scheme are complex and at times paradoxical, a movement "to the right," for instance, denoting possibly a leftward revolution when the full cycle is traced along the circumference and vice versa. In Dante, clockwise and counterclockwise movements are not always, respectively, sunwise (positive) or witherwise (negative), since directions in the northern and southern hemispheres are reversed. Symbolic value therefore depends on perspective and can vary considerably with changes of perspective. Blake generally abides by traditional directional significations, although the referent symbolized by either a positive (rightward) or negative (leftward) revolution is often quite unconventional. Jehovah and Newton, for example, are linked primarily with the sinister spiral. In Descartes's dream-vision, opposite directions stand for opposing choices or dispositions, and right and left have the explicit moral connotations of good and evil. It is in the context of just such a moral conflict that his symbolic *tourbillon* appears, like an eddy spawned by cross currents of a turbulent flux.

When compared with the vortical imagery studied in Part One, it is clear that a rich symbolic polyvalence continues to inhere in the vortex image, encompassing, once again, such antinomies as the one and the many, stasis and movement, order and disorder, creation and destruction, repetition and progression, centripetality and centrifugality, and so forth. There are, nonetheless, new associations that demonstrate the symbol's capacity to evolve in time, acquiring new meanings, obscuring older ones no longer relevant without altogether discarding them.

It would be impossible to recapitulate all of these nascent connotations, since they are too numerous and complex, but singling out one from each of the three authors examined in this part should give a sense of their originality and diversity and provide an appropriate comparison upon which to conclude. I begin with Dante's breathtaking evocation of a brilliantly choreographed celestial dance: whirling spirit-lights ardent with divine love, gyrating independently while others joyfully reel in great communal wheels, and all the redeemed souls swept along in paradise's vortical ecstasy. In Descartes, a dizzying creative vertigo, an exhilaration he terms "enthusiasm," linked to the experience of poetic inspiration and derived from profound insights achieved after scaling the intellectual Olympian heights. Finally, Blake's epiphanal vortex of transcendent poetic imagination, which Descartes's vertiginous enthusiasm seems uncannily to anticipate, whereby the "traveller thro Eternity," the poet—Milton, Blake, Dante, the young Descartes—passes, by means of divine inspiration and imagination, into the vortex that is each thing, seeking and gaining a perspective on its perspective, getting behind, outside appearances, spiraling through the vortical center to hidden truth beyond.

Part III

Threshold of the Unknown

Descents into Poe's Maelstrom

8

I think it altogether appropriate, in this last division of my inquiry, to consider Edgar Allan Poe a symbolist along with Rimbaud and Mallarmé, even though he did not participate directly in the French symbolist movement of the late nineteenth century.[1] The same could actually be said of Rimbaud, who was at best indirectly connected, and yet he is often referred to as a symbolist poet. But then aren't all poets symbolists? In the broadest sense, the answer of course is yes. I nevertheless mean to identify a specific nineteenth-century symbolist episteme by my grouping, with aims and ideas quite different from, for example, the authors of the Christian ethos we have just examined.

The term "Christian" is the key to this difference. For Dante, Descartes, and Blake, the Christian God presiding over his providential design is the transcendent signified toward which immanent signifiers of the phenomenal world point, and all three conceive of themselves as seers with a mission to penetrate through appearances to divine truths. The later symbolists retain the identity of the *voyant,* but the vision itself, while still the glimpse of a hidden ideal, is no longer specifically Christian. Shrouded in mystery, it may be represented as a distant, evanescent star, like Poe's "Al Aaraaf," or, as in the case of Rimbaud, religious terms—*Eternité, Mystique, Illuminations*—may evoke an aura of the sacred, even if the poet's insights are apt to be iconoclastic and profane.

Since the object of mystical discovery is uncertain or unknown, the transcendent act itself, while exhilarating and illuminating, is also fraught with doubt and trepidation, as though—to borrow a revealing and appropriate image from Baudelaire—the poet is peering vertiginously into a gaping abyss. "Le Gouffre" does effectively capture the terrifying quality of the newer symbolist breakthrough, freed of traditional and familiar associations and, hence, threateningly unspecified: no beatific apprehension of divinity, but rather the image of "un grand trou / Tout plein de vague horreur, menant on ne sait où"[2] ["a great hole / Full of vague horror, leading one knows not where"].

Already the image of Baudelaire's *gouffre* seems to convey us to the brink of Poe's maelstrom, which may have inspired it. It is of course to Pascal that Baudelaire alludes in his sonnet, but the more immediate influential source would be Poe, whose riveting encounters with deadly water chasms Baudelaire meticulously and compellingly rendered into French. While Baudelaire's abyss is not necessarily a maelstrom, Poe's maelstrom is quite obviously an abyss, and one that often quite explicitly manifests spiro-helical or vortical characteristics. Richard Wilbur, in a lecture commemorating Poe's sesquicentennial year, mentions "the recurrence of the *spiral* or *vortex*" as one of the author's most pervasive symbols, and he outlines its incidence in the tales:

> In *Ms. Found in a Bottle,* the story ends with a plunge into a whirlpool; the *Descent into the Maelström* also concludes in a watery vortex; the house of Usher, just before it plunges into the tarn, is swaddled in a whirlwind; the hero of *Metzengerstein,* Poe's first published story, perishes in "a whirlwind of chaotic fire"; and at the close of *King Pest,* Hugh Tarpaulin is cast into a puncheon of ale and disappears "amid a whirlpool of foam." That Poe offers us so many spirals or vortices in his fiction, and that they should always appear at the same terminal point in their respective narratives, is a strong indication that the spiral had some symbolic value for Poe. And it did.[3]

Wilbur does offer his own reading of the symbol, which I shall defer specifying as only seems appropriate, until after we have had a chance to study the first two stories, in which the image receives its most significant and elaborate development.

In "MS. Found in a Bottle," the earlier of the two (published in 1833), there are three successive incidents of violent and destructive turbulence. The narration of this gripping adventure traces at once an expansion of the vortical image and a progressive disintegration of the narrator's highly rationalistic "habits of rigid thought,"[4] as the vessel in which he is swept along descends ominously southward. The transformation from the calm order of his "contemplative turn of mind" or what he calls "the aridity of my genius," at the outset, to the dislocated, dizzying phantasmagoria of the later episodes is represented literally by a breaking up of the narrative itself, just over halfway through, into increasingly desperate and disjointed thought-fragments. In this way, the turbulence of the story's central vortex symbol disrupts both the events recounted and the recounting of events.

Properties of vorticity penetrate the text in many other ways. One recurring characteristic we have noted in various contexts throughout this study is the opposition of antinomies that creates conditions conducive to the production of a whirl. Poe represents this polarized state on several levels. Halliburton points, for example, to "the peculiar to-and-fro quality, as of a pendulum swinging,"[5] that characterizes the author's descriptive technique.[6] The up-and-down alternation between crest and trough of wave and whirlpool further amplifies the paradigm of conflicting currents in which the forlorn vessel and narrator seem, both literally and figuratively, to have been cast adrift.

The juxtaposition of stasis and violent activity is an opposition inherent in vorticity that Poe carefully works into his depiction of the first vortex event. Prior to the encounter, the narrator describes an eerie calm in which the air "became intolerably hot, and was loaded with spiral exhalations similar to those arising from heated iron" (II, 3). The stillness known to precede a hurricane is in this case associated with the whirlpool-tempest, and its appearance is presaged symbolically by the paradigmatic spirals. The narrator's "full presentiment of evil" is quickly realized when "a loud, humming noise, like that occasioned by the rapid revolution of a mill-wheel," heralds the storm's fierce and sudden onslaught. Caught unawares, the entire crew, save himself and "an old Swede," perish as they sleep, "so terrific, beyond the wildest imagination, was the whirlpool of mountainous and foaming ocean within which we were engulfed" (II, 4). Here, the vortex is literally a natural destructive force, but figuratively, just as the stifling lull of the Malaysian sea was blasted by the tempest, so too have the narrator's ease and "arid" rationality been unexpectedly jolted by an evil, if yet incomprehensible, turbulence.

As the survivors drift toward the next catastrophe, the concept of descent is linked with that of evil in a manner that recalls Dante's journey to the underworld.[7] Moving ever faster and "farther to the southward than any previous navigators" (II, 6), they enter a dismal region of "pitchy darkness," where, according to the narrator, "Eternal night continued to envelop us" and "All around were horror, and thick gloom and a black sweltering desert of ebony." Dante's descent was also a southerly one into gloomy, eternal darkness, as we have seen, and the narrator's statement that "I could not help feeling the utter hopelessness of hope itself" might well have been spoken by the Italian pilgrim-poet in his despair. The parallel becomes all the more explicit when, caught in a whirlpool-abyss, the narrator observes that "at times [we] became dizzy with the velocity of our descent into some watery hell" (II, 6–7).[8]

The episode is perhaps Poe's most fanciful depiction of dialectic associated with a vertiginous descent. As before, the turbulence is destructive—the Swede and the vessel perish—and the narrator is once again miraculously spared, but this time due to a most astonishing peripeteia. A gigantic "ghost-ship" appears suddenly at the rim of the abyss into which the narrator's vessel has already deeply plunged. When, after the two ships collide, he is hurled violently from his doomed craft onto the rigging of "the stranger" (II, 8), it is soon apparent that the shift is not merely one of place (from vessel to vessel), but also a transition from one "dimension" into another, from "the world" into an ethereal "anti-world," as if he had passed beyond the "event horizon" of a black hole through a "wormhole" into a bizarre inverse universe. There are many details in the series of discrete thought-fragments (now supplanting the previously continuous storyline) that betray the antithetical, irrational nature of this other world, and they usually involve circumstances contrary to the narrator's experience of the "real world" he has just left or conditions impossibly contradictory in and of themselves.

The specifically negative relation this strange new "anti-reality" bears to "reality" is evident in the narrator's statement, concerning the ship, that "What she *is not,* I can easily perceive—what she *is* I fear it is impossible to say" (II, 10). This negative reversal also applies to his own physical presence (now an absence), since the ghost-ship's crew, from whom he at first attempts to hide, pays him no heed, as if to them he does not exist: "Incomprehensible men! Wrapped up in meditations of a kind which I cannot divine, they pass me by unnoticed. Concealment is utter folly on my part, for the people *will not* see" (II, 9). His feeling, too, of "a sensation which will admit of no analysis . . . so utterly novel" (II, 9) contrasts sharply with the habits of ratiocinative "methodizing" and self-conscious *ennui* that characterize him at the outset.

As for contradictions within the contradiction of the ghost-ship's anti-world, it is precisely because conditions are the reverse of those in the real world (where contraries are by definition mutually exclusive) that oppositions can coexist in ways that seem at once "normal" and supernatural. Hence, the aged and infirm ghost-captain's "manner" is "a wild mixture of the peevishness of second childhood, and the solemn dignity of a God" (II, 9), an incompatible union of youth and old age, imperfection and divinity that could only exist or be imagined outside of time, beyond the threshold of an event horizon. In another example, the narrator suspects "the operation of ungoverned Chance" (II, 10) when he involuntarily spells out the word "Discovery" with seemingly random daubs of tar on a studding-sail. At the same time, his conclusion that "I must suppose the ship to be within the influence of some strong current, or impetuous undertow" suggests, to the contrary, that a directing force like the "Polar Spirit" in Coleridge's "Mariner" is controlling events, and this force is a hellishly evil one, as the allusion to "demons of the deep" that "rear their heads above us" (II, 12) clearly implies.

The demons simile once again evokes the anti-world of Dante's infernal vortex, where contradiction is similarly the norm, and the analogy is unmistakable in the description of the last, terrifying vortical encounter. To begin with, references to eternal darkness, chaos, and towering walls of ice describe a grim and forbidding landscape very much like the frozen desolation of nether hell's Lake Cocytus:

> All in the immediate vicinity of the ship is the blackness of eternal night, and a chaos of foamless water; but, about a league on either side of us, may be seen, indistinctly and at intervals, stupendous ramparts of ice, towering away into the desolate sky, and looking like walls of the universe. (II, 13–14)

Also, in the harrowing closing image describing the ship's dire descent, the whirlpool's immense "amphitheater" of centripetally narrowing concentric circles incorporates congruently the structure of Dante's inferno:

> —Oh, horror upon horror! the ice opens suddenly to the right, and to the left, and we are whirling dizzily, in immense concentric circles, round and round the borders of a gigantic amphitheather, the summit of whose walls is lost in the darkness and the distance. But little time will be left me to ponder upon my destiny—

> the circles rapidly grow small—we are plunging madly within the grasp of the whirlpool—and amid a roaring, and bellowing, and thundering of ocean and of tempest, the ship is quivering, oh God! and—going down. (II, 14–15)

There can be no doubt about the destructive nature of Poe's maelstroms, which appear and intensify amidst ever-increasing conflictual cross currents and antinomies, as is so often the case with the vortex, be it phenomenon of nature or symbol. And the Dantesque correspondence heightens the atmosphere of evil, hopelessness, and death.

Still, the author has taken pains to imbue the experience of the descent with symbolic ambivalence. When the narrator exclaims, "We are surely doomed to hover continually upon the brink of Eternity, without taking a final plunge into the abyss" (II, 12), not only do we recognize the dramatic slowing down of time that modern theory posits for an object caught in a black hole's cosmic vortex (see Appendix) and that Poe has so presciently intuited, but it also seems as if the narrator would very much like to take the plunge. Then, near the end of the tale, the "imp of the perverse" gets the best of him:

> Yet a curiosity to penetrate the mysteries of these awful regions, predominates even over my despair, and will reconcile me to the most hideous aspect of death. It is evident that we are hurrying onwards to some exciting knowledge—some never-to-be-imparted secret, whose attainment is destruction. (II, 14)

Whether overcome by some sort of thrill-seeking death instinct or yielding to impulses of the "new entity" that now possesses his soul, the narrator's curiosity impels him toward a wonderful and mysterious encounter with essential truth, which, as the story's epigraph implies,[9] awaits discovery in the terrifying depths of the abyss, a secret knowledge whose attainment means destruction, not of being itself, but dissimulations of being, so that the self, unburdened of its waking hypocrisies, can spiral beyond mere consciousness of the real into a dream zone of transcendent visions and intuitions. The ghost-ship's voyage is thus a descent into a time warp of pure imagination, beyond perception and reason (he is invisible, his acts involuntary), beyond language (utterly useless and incomprehensible), toward some primal matrix of unmediated meaning, where signifiers float freely,[10] negations are themselves negated and all vain human dissimulations and cover-ups *dis*-covered by the maelstrom's destructively cleansing whirl.

In his later story, "A Descent into the Maelström" (1841),[11] Poe takes up the vortex symbol once again, making it this time the unique focus of the tale, as the title suggests. Many themes from the earlier work recur, but the structures and outcomes of the two are quite different. Instead of a "manuscript in a bottle"[12] surviving to record the experience, an eerie old sailor with ghostly white hair—reminiscent of Coleridge's ancient mariner, as many have observed—relates with similar obsession the chilling account of his descent into the "Moskoe-ström" (the local name for the great Norwegian whirlpool) to a curious, if nonetheless reluctantly obliging interlocutor, who, like Coleridge's wedding guest, is the story's narrator. That the old seafarer lives to tell his story is evidence at the outset of its fortunate ending, as opposed to that of the earlier story from which nothing remains but the bottle's posthumous message.

Whereas the earlier tale chronicles three sequential episodes of turbulence, the "Descent" records two, and they take place, as it were, simultaneously, since an actually occurring incidence of the maelstrom serves as a dramatic backdrop to the old sailor's recall of his perilous descent some three years prior. The concentric structure of the story's symbolic maelstrom-abyss is thus mirrored paradigmatically by the narrative *mise en abyme* of tale within tale. This enables the author to create temporal ambivalence by interfacing past and present vortex events "contrapuntally," compared to the more linear juxtaposition in "MS." of the opening's present real world with, ultimately, a phantasmagorical anti-world situated in some timeless past or, as the narrator puts it, "imbued with the spirit of Eld" (II, 13). Generally, the atmosphere and dénouement of the "Descent" are less incredible and fantastic than in "MS.," even though the narrative does seem to spiral progressively deeper into the told story's past toward the threshold of death and an unsettling glimpse of the unknown.

The depiction of the presently occurring vortex that dominates the first part of the story is quite realistic, and the narrator, although observing the maelstrom from the dizzying heights of Mt. Helseggen with mixed feelings of rapture and terror, provides a rather matter-of-fact description of the whirlpool forming before his eyes and relates in detail various accounts of its history and genesis. He even quotes from the *Encyclopedia Brittanica,* one of Poe's actual sources, although these passages had in fact been "lifted" by the encyclopedists—as Peithman notes (96) – from Bishop Pontoppidan's study, *The Natural History of Norway.*[13]

In many ways, the factual details serve to heighten suspense by prolonging the reader's anticipation of the old sailor's narration of "an event such as never happened before to mortal man" (II, 225) to which he tantalizingly alludes at the very outset of the story. Intermingled with the "facts" are somber descriptive elements that contribute to an atmosphere of pervasive gloom and impending disaster. For example, while watching the maelstrom form from "gigantic and innumerable vortices," the narrator describes them as "all whirling and plunging on to the eastward with a rapidity which water never elsewhere assumes except in precipitous descents" (II, 228). Here, it appears that the vortex whirls in the ominous witherwise (eastward, leftward) direction, although we need to know what direction he is facing and whether he is speaking of the top or bottom of the arc to be sure. An infernal Dantesque allusion further intensifies the dismal ambience when the narrator depicts himself "Looking down from this pinnacle upon the howling Phlegethon below" (II, 231), his own "giddy" perspective from the cliff's edge foreshadowing the seafarer's view from "on the brink of the maelstrom" down into the gaping hole of the abyss beneath him.

The mariner's remark early on that his hair was changed "from a jetty black to white" in "less than a single day" (II, 225) is clearly the most dramatic evidence of the trauma he has experienced, but it also initiates a stark symbolism of polar antithesis echoed, in the first part of the tale, by references to vortex-producing cross currents and alternations between extremes. The narrator notes, for example, "a thousand conflicting channels" (II, 228) as the maelstrom emerges, and the encyclopedia article refers to trees "whirled *to and fro*" by the stream, which, it claims, "is regulated by the *flux and reflux* of the sea – it being constantly *high and low* water every six hours" (II, 231, emphasis added). But the narrator and the mariner both doubt the encyclopedia's theory that the vortices of the region " 'have no other cause than the collision of waves *rising and falling*, at *flux and reflux*, against a ridge of rocks and shelves' . . . for, however conclusive on paper, it becomes altogether unintelligible, and even absurd, amid the thunder of the abyss" (II, 232, emphasis added). All the while, in yet another example of dialectical ambivalence, the narrator, like his predecessor in "MS.," confesses to conflicting feelings of terror and awe in the face of this titanic natural phenomenon, when he proclaims the failure of Ramus's scholarly account to "impart the faintest conception either of the magnificence, or the horror of the scene—or of the *novel* which confounds the beholder" (II, 229).

The structural dualisms and contradictions continue in the old sailor's retrospective reconstruction of his "descent into the maelstrom," which Poe has so masterfully prepared. The suspense is immediately intensified by a sudden cessation of the wind (cf. the first vortex event of "MS."), leaving the sailor and his brothers stranded or, as he says, "dead becalmed" (II, 235) in the middle of the channel, just when they should be scurrying to port in order to avoid "the Ström." Stillness soon gives way to fury, and the hurricane that strikes suddenly drives the paltry "smack" toward the whirlpool, the approach to which is described in terms of a disorienting alternation between crests and troughs:

> Presently a gigantic sea happened to take us right under the counter and bore us with it as it rose—up—up—as if into the sky. I would not have believed that any wave could rise so high. And then down we came with a sweep, a slide and a plunge, that made me feel sick and dizzy, as if I was falling from some lofty mountain-top in a dream. (II, 238)

This last allusion also reflects the narrator's "present" position perched on the towering cliffs, as he listens to the mariner's yarn, and reinforces the parallels Poe has carefully drawn between the two now simultaneously occurring vortex episodes.

Like the narrator in the first part (and also like the speaker of "MS."), the old sailor expresses mixed emotions as he undergoes the experience, but in this case both his feelings and his fate take an unexpected turn of events. A possible harbinger of the story's happy outcome may be the maelstrom's rightward, sunwise (and hence, symbolically propitious) circumvolution. This is made clear, first, when the mariner says of the smack that "Her starboard side was next the whirl, and on the larboard [left] arose the world of ocean we had left" (II, 239) and again, later, when he recounts, "we gave a wild lurch to starboard, and rushed headlong into the abyss" (II, 241). In the meantime, his decision "to hope no more" (II, 239) has somehow diminished his sense of terror, and he has begun to imagine "how magnificent a thing it was to die in such a manner." Then, perhaps because of the vertiginous, disorienting effect of the revolutions, a curious mixture of "the imp of the perverse" and an almost scientific inquisitiveness overcomes him: "After a little while I became possessed with the keenest curiosity about the whirl itself. I positively felt a *wish* to explore its depths, even at the sacrifice I was going to make" (II, 240).

In the end, it is this strange blend of contradictory emotions—"sensations of awe, horror and admiration" (II, 242) and even "amusement" (II, 243)—yielding to "the dawn of a more exciting *hope*" (II, 244) that ultimately enables him to survive. During the long moments of his descent, he notices with an abnormally acute interest that small, cylindrical objects are absorbed by the whirl less rapidly than others. For once, the typically ratiocinative penchant of Poe's narrators, heightened in this case by a kind of delirious fascination with the sheer power and beauty of the vortex, pays off. Recalling, with an incredibly dispassionate lucidity, the findings of Archimedes—the reference is a specious invention of Poe's—that "a cylinder, swimming in a vortex, offered more resistance to its suction, and was drawn in with greater difficulty than an equally bulky body, of any form whatever" (II, 245), the mariner "precipitates" himself into the raging surf lashed to a barrel and thus retards his descent long enough to outlast the Moskoe-ström.

In the debate over whether it is his rational or aesthetic sense ("Taste") that "saves him," Sweeney's reading of the narrative as an illustration of "Poe's thesis that, as regards the chaotic mysteries of Nature, aesthetic intuition is far more important than science"[14] seems to me to overstate the case. The sailor's feeling of wonder in the face of a terrible beauty does help him overcome his initial fear, so that he can *act* upon his impulse to take a calculated risk, but it is his ratiocinative tendency that motivates him to calculate the risk in the first place. Shulman's attribution of his success to "the counter forces of reason and imagination"[15] strikes me as a more accurate assessment. Hence, Poe's "thesis" would seem rather to be that mastering terror, resourcefully directing its enormous charge of emotional and psychic energy toward creative, constructive ends in harmony with the forces of nature can have amazingly salutary results. Blind self-interest, lack of imagination, and unthinking surrender to panic and fright, the treacherous elder brother's response to calamity, represent, on the other hand, a rigidity of spirit that resists and even repudiates nature. Such lack of resilience cannot withstand the maelstrom's superior force and will be whirled to destruction, as events ultimately prove.

Although the tale's unexpected reversal may at first seem surprising, it is really quite consistent with Poe's portrayal of the vortex symbol. Entering a vortex, in Poe's universe, means passing into a dimension of pure imagination in which routines, complacencies, and habitual expectations break down. One may cross into an anti-world in which one sees and yet is unseen, speaks but is unheard, or to cite the case in point, it may turn out that the one who has the surest grip and seems most safe perishes, while the one who takes the greatest risk and seems to doom himself by plunging headlong into chaos survives.[16]

The oftentimes "irrational" occurrences in Poe's vortex are actually not unlike those of the dream state, and Wilbur claims that "What the spiral [vortex] invariably represents in any tale of Poe's is the loss of consciousness, and the descent of the mind into sleep" (257). Marie Bonaparte, in her "Psycho-Analytic Interpretation" of Poe's life and works, focuses on the latent dream content of Poe's symbolism, and she detects in each of his whirlpools "a version of the return-to-the-womb phantasy."[17] Apropos of the "Descent," she writes,

> The hero escapes the fate to which his brother falls victim, in the same way that Poe survived his brother Henry . . . [succeeding], as it were, in touching bottom, in reaching those innermost uterine depths where the foetus once lay, bathed in those amniotic waters which are of the few vestiges of the parent ocean from which, phylogenetically, we have all sprung. (352)

In an earlier chapter, spiro-helical and vortical configurations were linked to the themes of feminine sexuality and birth, and Poe's interest in the symbol may indeed stem from a neurotic obsession. There does not seem to be much in the text, though, beyond the vortex symbol itself and an unproven identification of Poe with his characters, to corroborate Bonaparte's psychoanalytical interpretation.

Halliburton, in his phenomenological reading, emphasizes the spatial and temporal dimensions:

> Now a vortex is as much a space-*toward* as a space-between: in a vortex one is "on the way." But there are differences. The first difference is that one's descent in the vortex is gradual, notwithstanding the velocity of motion. One cannot go down very fast when one is also going round and round. In this connection we are reminded of Poe's concern with gradation, the shades of development by which one condition accedes to another. All that must be said of that subject here is that one cannot breach the gulf beyond

> all at once: *Novelty* is preceded by stages of the merely novel. The second point I want to make is that the vortex—the pure spiral movement—is the most equivocal of motions. It is a descent that resists descent, a movement that twists away from itself only to be twisted back to itself *by* itself. It is as near to non-movement as any movement can be. (249–50)

Herein lies the paradox of Poe's vortex. It is, on the one hand, a threshold, a transitional phase, a "space-toward" that links separate dimensions: the "magnificent rainbow" of the "Descent," "like that narrow and tottering bridge which Mussulmen say is the only pathway between Time and Eternity" (II, 243). And yet, because the spiro-helical descent is so slow, "as near to non-movement as any movement can be" or, in Serres's terminology, a "circum-stance," it is a state of "suspension" (in a dimension of its own), "a kind of indefinite present" and thus conducive to the effect of prolonged "suspense" that Poe so eagerly seeks to realize in his depiction of the narrator-victim's "eternal moment of terror" (HP 250).

Poe's vortex, like Blake's, does nevertheless imply an eventual passage into an unknown realm beyond—be it "novelty," "eternity," "death," "unity," or the "truth" at the bottom of Democritus's well, the depth of which is far greater than imagined, according to the epigraph of the "Descent," just as our feeble conceptions of God's ways are in no way "commensurate to the vastness, profundity, and unsearchableness of His works" (II, 225). Our apprehension of the everyday world outside the transformational vortex, like that of a person "who from the top of Aetna casts his eyes leisurely around, is affected chiefly by the *extent* and *diversity* of the scene. Only by a rapid whirling on his heel could he hope to comprehend the panorama in the sublimity of its *oneness*."[18]

In order to attain the "*individuality* of impression" (XVI, 187) that constitutes a vision of the "Original Unity of the First Thing" (XVI, 185), i.e., Poe's "Godhead," "we require something like a mental gyration on the heel. We need so rapid a revolution of all things about the central point of sight that, while the minutiae vanish altogether, even the more conspicuous objects become blended into one" (XVI, 187). As was the case for Descartes, swept up in a whirlwind while scaling the heights of his own Olympian Aetna, the visionary breakthrough for Poe is the result of a dizzying, disorienting experience, occurring in the context of conflict and crisis—"getting caught in a maelstrom," as it were—by means of which all distracting multiplicities are blurred together by the whirl into a thrilling and sublime glimpse not only of nature's "terrific grandeur" (II, 242), but of the essential unity underlying all mere appearances.

"Tourbillons de Lumière": Rimbaud's Illuminating Vortices

9

The image of Poe's narrators peering from on edge into the turbulence of the abyss, overcome by a morbid curiosity about the unknown beyond and by a sense of fascination with the very "novelty" of the experience, seems already to describe the situation of Rimbaud's "poet-seer." Repeatedly in his letters, the *enfant terrible* of French symbolism affirms, in his enthusiastic elaboration of a poetic aesthetics, that "Il s'agit d'arriver à l'inconnu" ["It's a question of reaching the unknown"],[1] and echoing both Baudelaire and Poe, he exhorts, "demandons aux *poètes* du *nouveau* ["let's ask poets for the new"].[2]

So radical in fact are Rimbaud's innovations that any critical venture into his bizarre symbolic universe must inevitably address what Breton has called "the Rimbaud problem," resulting from an undisguised assault upon all conventions (even at the basic semantic and syntactic level of words and phrases) that calls into question the most essential hermeneutical assumptions and casts doubt on the relevance or even the feasibility of interpretation in his case. The poet has, after all, challenged his readers and critics rather mockingly to solve his poetic puzzles ("trouvez Hortense" ["find Hortense"]),[3] while reserving to himself alone privileged access to these meanings ("J'ai seul la clef de cette parade sauvage" ["I alone have the key to this savage parade"]).[4] Todorov goes so far as to assert, in the case of the *Illuminations,* that the very lack of any accessible meaning is itself the intended meaning of these poems: "To want to discover what they mean is to strip them of their essential message, which is precisely the affirmation of an impossibility to identify the referent and understand its meaning."[5]

I am not so sure these texts are utterly impenetrable and hermetic. Nonetheless, the lack of referents in reality for many word images (a purely "poetic" function of language that Ross Chambers terms "non-mimetic"),[6] complicated by a breakdown in the linguistic sign itself, that is, signifiers corresponding frequently to incongruous or outright nonsensical signifieds, do pose a fundamental obstacle to any clear understanding of a poem's meaning. A second important characteristic of Rimbaud's poetics, recognized by many, is an emphasis upon synecdochal fragments without any explicit reference to the whole of which they are parts: "The text names the parts, but they are not there 'for the whole'; they are rather 'parts without the whole'" (TC 249). Consequently, each word, phrase, or image is often an autonomous "island," which may or may not relate syntagmatically to the words surrounding it, with the result that "the only relationships between events or between sentences that Rimbaud cultivates are of co-presence . . . of pure spatial and temporal co-presence" (TC 246–47).

The critic's function, then, if not rendered altogether moot, has changed radically with Rimbaud, since before attempting to understand the meaning of the poet's symbolism, one must first determine what symbols are actually present and how they "correspond" in the complex fabric of ostensibly discontinuous and fractured image fragments. To this end, the text-centered "paradigmatic criticism" described by Todorov[7] seems to me to be an appropriate and promising one in the case of Rimbaud and of similarly hermetic symbolist poets. Accordingly, my aim, as far as the *tourbillon* or vortex paradigm is concerned, must first be to disengage it by calling attention to the structures that, taken together, denote its presence in a poem and then to come to terms with its symbolic significance in the poet's *œuvre*.

The vortical image is at first rather incidental in the poems, and "Le Bateau ivre" ["The Drunken Boat"] (R 128–31) is the first one in which it figures prominently at all, although it seems primarily to represent part of the tempestuous marine *décor*, serving to intensify the turbulence of the fantasy voyage. The poet's embarking upon a watery descent at the very outset and his insistence upon downward movement throughout—the verb *descendre* recurs at least four times—strongly suggest a Poesque descent into the maelstrom, and several explicit vortex structures incorporate the image: "les trombes" ["waterspouts"] (l. 29) that the narrator "knows," the abstract allusion to "les lointains vers les gouffres cataractant" ["distances cataracting toward abysses"] (l. 52), the vivid clash of contraries in the oxymoron "Les cieux ultramarins aux ardents entonnoirs" ["ultramarine skies with burning funnels"] (l. 80), and the direct reference to "les Maelstroms épais" ["thick maelstroms"] (l. 82) that made the poet tremble. Waterspouts, abysses, fiery funnels,[8] and maelstroms—the plural in each case underscores both the ubiquity and nonspecific nature of the symbol—combine to portray a storm-tossed, treacherous seascape, but one so fantastic that it could only exist in a dream-vision, which "Le Bateau ivre" certainly is. In fact, the vortical whirl in Rimbaud seems initially to denote the dream state, as is clear in his very first poem, "Les Etrennes des orphelins" ["The Orphans' Gifts"] (R 35–38), where the two children dream nostalgically of toys, candies, and jewels that begin to whirl about (*tourbillonner*) in a "sonorous dance."[9]

In Rimbaud's most vituperative political outburst, "Qu'est-ce pour nous, mon coeur" ["What do we care, my heart"] (R 171–72), a bitter, nihilistic reaction to the failure of the Commune, it is the dangerous, destructive quality of the whirlwind that appears to prevail, as the poet imagines and in fact vociferously advocates "les tourbillons de feu furieux" ["whirlwinds of furious fire"] (l. 13), i.e., a worldwide conflagration, not only to avenge the fallen *communards,* but also to overthrow all established orders. Recalling the image of "ardents entonnoirs," it is the fiery vortex that Rimbaud selects, this time to convey his vision of apocalyptic anarchy. The last verse—"Ce n'est rien! j'y suis! j'y suis toujours" ["It's nothing! I'm here! I'm still here!"]—may well suggest, as Suzanne Bernard claims, "the return to reality, the sobering awakening after the grandiose nightmare" (R 443), placing the vortex, once again, in the context of a dream.

But Rimbaud's most elaborate workings-out of the *tourbillon* occur in three "illuminations," including the two free-verse poems, "Marine" and "Mouvement," and the prose-poem "Mystique." In the first of these, disparate phenomena are related, not through simile or some other such analogous figure, but by means of the con-fusion effected by a vortical whirl.

MARINE

Les chars d'argent et de cuivre—
Les proues d'acier et d'argent—
Battent l'écume,—

Soulèvent les souches des ronces.
Les courants de la lande,
Et les ornières immenses du reflux,
Filent circulairement vers l'est,
Vers les piliers de la forêt, –
Vers les fûts de la jetée,
Dont l'angle est heurté par des tourbillons de lumière.

Chariots of silver and copper –
Prows of steel and silver –
Beat the foam, –
Raise up the stumps of bramble.
The currents of the moor,
And the immense ruts of the ebb tide,
Flow circularly toward the east,
Toward the pillars of the forest, –
Toward the boles of the jetty,
Whose angle is struck by vortices of light.

(R 287)

From a purely grammatical point of view,[10] a series of sentence fragments divide initially into two groups: subjects, of which there are four (ll. 1, 2, 5, 6), and predicates, of which there are three (ll. 3, 4, 7–10). The last of these includes three separate, although parallel, "destinations" introduced by the preposition *vers* and a final synthetic relative clause that fuses these elements by means of a referentially ambiguous *dont*.

The subjects are grouped in two pairs, and since each subject is plural, it can join independently with each of the predicates, which are plural too, but the predicates are also capable of being linked syntactically to any or even all of the subjects. This free association of subject-predicate is enhanced by the dashes at the end of verses and by the strict parallelism of the construction. Admittedly, the diachronic succession of fragments, when read in sequence, is ordered by the poet, but once seized by the mind, they are easily interchanged in the free flow of psychic associations and correspondences. The poet thus creates what might be termed a "dynamics of interchangeability," which lends itself to the vorticity first suggested by the predicate, "filent circulairement vers l'est."

The context in which turbulence is apt to be found, as we have repeatedly seen, is one of opposition, tension, and possible fusion, which in the case of "Marine," has just been demonstrated on the grammatical level. Thematically, these conditions are even more overt. An essential opposition between "landscape" and "seascape" runs throughout the poem, as many critics have observed, but instead of comparing them as discrete, if analogous, domains, Rimbaud mixes them by means of the syntactic free association just described and also by fusing maritime and bucolic synecdoches in a single subject or image, creating thereby the kind of nonreferential reality peculiar to his poems. The term for this synthetic procedure is "crossfade technique," according to Hugo Friedrich, who adds that "Marine" is the first modern example of it.[11]

Hence, the *chars* and *proues*, linked by the metallic qualifier *d'argent*, both "battent l'écume" and "soulèvent les souches des ronces." Here, the contrary actions of *battre* and *soulever* (opposing, as Plessen points out, "movement from high to low" to "movement from low to high" [23]) also suggest the properties of destruction and creation that are associated with the *tourbillon* and, in effect, synthesized by it. This fusion is further manifested in the two subjects that follow. Instead of "Les courants du reflux" ["The currents of the ebb tide"] and "les ornières immenses de la lande" ["the immense ruts of the moor"], as one might expect, the poet reverses the qualifiers to emphasize the turbulent union of "sea" and "land" and creates the following hybrids: "Les courants de la lande, / Et les ornières immenses du reflux."[12] Similarly, in the two directional phrases indicating a clash of cross currents,[13] the destinations are again expressed in terms of the sea/land dichotomy and again the qualifiers reversed. So instead of "les piliers de la jetée" ["the pillars of the jetty"] and "les fûts de la forêt" ["the boles of the forest"], the poet crosses the two to produce "les piliers de la forêt" and "les fûts de la jetée."

Finally, the collision of the currents yields the synthesizing *tourbillons* with which the poem culminates, just as adverse flows or tides induce whirlpools in nature, although Rimbaud's vortices are hardly natural phenomena but rather "vortices of light." Composed of neither water nor earth, as the sea/land confusion would imply, the poet again forsakes logic and reverses expectations. Remembering, however, that this poem is an "illumination," the image makes sense in terms of the work as a whole: the brilliance of poetic vision "illuminates" the unity underlying ostensibly disparate elements by breaking through appearances to discover profounder truths, just as a whirlpool pierces through the surface toward hidden regions of the deep.

If the dynamics of "Marine" are obviously vortical, since the *tourbillon* is explicitly named, the motion in "Mouvement," while again "turbulent," is evinced more subtly, and although the word *tourbillon* does not occur per se, its presence is indicated by the ensemble of vortical synecdoches that abound throughout the text. On the whole, "Mouvement" does seem to be a visionary poem whose "prophetic message" (GR 237) may well convey a "revelation concerning the ultimate destiny of humanity" (GR 234); the themes of scholarly research, scientific discovery, and progress are evoked. My purpose, though, is not to elaborate them in detail—others have—but to disengage the structure of the vortical context in which Rimbaud's prophetic vision occurs.

MOUVEMENT

Le mouvement de lacet sur la berge des chutes du fleuve,
Le gouffre à l'étambot,
La célérité de la rampe,
L'énorme passade du courant
Mènent par les lumières inouïes
Et la nouveauté chimique
Les voyageurs entourés des trombes du val
Et du strom.

Ce sont les conquérants du monde
Cherchant la fortune chimique personnelle;
Le sport et le comfort voyagent avec eux;
Ils emmènent l'éducation
Des races, des classes et des bêtes, sur ce vaisseau
Repos et vertige
A la lumière diluvienne,
Aux terribles soirs d'étude.

Car de la causerie parmi les appareils, le sang, les fleurs,
  le feu, les bijoux,
Des comptes agités à ce bord fuyard,
—On voit, roulant comme une digue au delà de la route
  hydraulique motrice,
Monstrueux, s'éclairant sans fin,—leur stock d'études;
Eux chassés dans l'extase harmonique,
Et l'héroïsme de la découverte.

Aux accidents atmosphériques les plus surprenants,
Un couple de jeunesse, s'isole sur l'arche,
—Est-ce ancienne sauvagerie qu'on pardonne?—
Et chante et se poste.

The swaying motion on the bank of the river falls,
The chasm at the sternpost,
The swiftness of the incline,
The enormous passing of the current
Conduct through unimaginable lights
And chemical newness
Voyagers surrounded by waterspouts of the valley
And the maelstrom.

They are the conquerors of the world
Seeking their personal chemical fortune;
Sport and comfort travel with them,
They take the education
Of races, classes and beasts, on this vessel
Repose and vertigo
To diluvian light,
To terrible evenings of study.

For from the talk among the apparatus, blood, flowers, fire, jewels,
From the agitated accounts on this fugitive deck,
—One sees, rolling like a dyke beyond the hydraulic motor road
Monstrous, lighting up endlessly,—their stock of studies;
Themselves driven into harmonic ecstasy,
And the heroism of discovery.

Amid the most surprising atmospheric accidents,
A youthful couple, stands isolated on the arch (ark),
—Is it ancient wildness that one pardons?—
And sings and stands guard.

(R 304–5)

First, the network of conflictual cross currents and alternations between polar antinomies that form the matrix from which the synthetic vortex issues must be established. Since it is a question of a vessel at sea, as in "Le Bateau ivre," and a series of sea images, as in "Marine," the dynamics are principally, although not exclusively, hydraulic (as the allusion to "la route hydraulique motrice" of the third strophe suggests). "Le mouvement de lacet" of the first verse immediately sets the dialectical pattern of back and forth to which the subsequent antinomies conform. "This zigzag theme" is then echoed by the reference to "l'énorme passade" in line 4, according to Kittang, who cites equestrian and theatrical meanings and even one from falconry to support his reading.[14]

Among the series of polar opposites between which the poem zigzags, perhaps the most important is the dual orientation toward past and future from the present perspective of the "actual" sea voyage. On the one hand, the poet alludes retrospectively to Noah's ark, the couple isolated "sur l'arche" recalling mankind's "ancienne sauvagerie" and captured, as it were, in "la lumière diluvienne" of the epoch of the Flood. In my reading, the true "extase harmonique" is meant to be associated with this original natural state, after the earth was cleansed. On the other hand, the perspective is futuristic, focusing on the modern, educated man, "la nouveauté chimique" with which he is obsessed and the new scientific "héroïsme de la découverte" that it has produced. Here, I disagree with those critics who portray a Rimbaud celebrating the scientific progress of the age. On the contrary, the tone seems quite sarcastic, likening these "conquérants du monde / Cherchant la fortune chimique personnelle" (in drug-induced *paradis artificiels?*) to greedy merchants hawking their wares ("leur stock d'études"), which are no less than "monstrueux." Rimbaud's iconoclasm and preference of "la bohème" to the dehumanizing "appareils" and extravagant luxuries ("bijoux") of the age's bourgeois lifestyle would seem to indicate such an ironic reading.

This temporal opposition is mirrored by other antinomies. There is the reference to "repos et vertige," i.e., the alternation between extremes of rest and frenzied activity that characterizes the dynamics of progress through history and the very phenomenon of "movement" itself that we experience and take for granted and that is "illuminated" by poetic insight. Also, as in "Marine," the correspondence between land and sea is portrayed in the conjunction of "berge" and "chutes du fleuve" in the opening image, as well as the earthy and watery "trombes du val / Et du strom" with which the first strophe closes. Lastly, the "couple de jeunesse" at the end of the poem literally personify the theme of polar dualism in a truly dialectical manner, since the two of them are nonetheless "*un* couple."

Another vortical synecdoche is the idea of "declination" or "inclination" that Serres refers to so frequently in his study of Lucretius. The conical form of the vortex, like the "V" of the first letter, implies a slanting descent, as well as, ultimately, the angle of intersecting oblique vectors. In the first strophe, which like "Marine" consists principally of a series of substantives (introduced matter-of-factly by the definite article), declination and descent are evoked in the following image cluster: the sloping sides of the *berge* and *val* (again mirrored in the initial letter), the incline suggested by the word *rampe* (a stair's slanting handrail?), and the descent unequivocally indicated by the *chutes du fleuve*.

Finally, lest there be any doubt that the vortex figures here, especially in the first stanza, "Le gouffre à l'étambot" describes the whirlpool that occurs naturally at the stern in the ship's wake. The "trombes du val / Et du strom" are cyclones that "surround" the voyagers, as if they were themselves caught in a *tourbillon,* which seems quite definitely implied by the poet's choice of the foreign *strom,* associated immediately with the current of a "maelstrom." And the vortical whirl is further denoted by the *vertige*[15] of the second stanza, which is echoed in the "agitated" and "rolling" motions of the third, culminating in "l'extase harmonique" of heroic discovery, but not really that of the new chemists, who, as I have argued, are ecstatic only about their personal fortunes and, as such, are objects of the poet's scorn, but rather the vertiginous "dis-covery" of poetic insight, to which the poem itself testifies in its uncovering (illuminating) of the essential structure of "movement."

The pivotal symbol is the couple isolated on the archway, who occupy the focal point of the visionary vortex and embody its dialectical (antithetical and synthetic) properties. They are two, who by uniting produce another, and so on in the chain of dialectical interactions that "motivate" (i.e., "give movement to") the process of human propagation. Like the "arch" on which they stand, a bridge between opposite sides, they also link past and future, reminding the poet, on the one hand, of passengers on Noah's "ark" and of mankind's salvation after the Flood and, on the other hand, of the generations yet to come that will spring from them. In such a manner, the poem seems to conclude optimistically, the poet perceiving in their youth and innocence the promise of a modern deliverance from the sterile, egotistical scientism and bourgeois materialism of the other voyagers and, by extension, of the age.

Although it seems, at this point, that Rimbaud's vortical symbolism emanates principally from sea images and themes, it is by no means confined to them. In "Mystique," there are certain maritime allusions, but the scene is more ostensibly a terrestrial one and the vorticity more abstract. Moreover, the mystical, visionary character of the poem, announced in the title, and the resultant supernatural quality of the images pose even greater obstacles to critical analysis, although by concentrating on the network of underlying structures, the vortex symbol, which is not stated outright, can be rendered explicit.

Perhaps more than any other "illumination," "Mystique" is the type of "gravure colorée" (colored plate) that Verlaine, in his preface to the 1886 edition, defines as the intended meaning of Rimbaud's title. Confronting it, the reader literally beholds a painted tableau, the top, bottom, and sides of which the poet meticulously describes in the course of the poem's four parts. In effect, it is a window into a kind of noumenal world beyond that of phenomenal appearances, one outside of reason and logic and essentially nonreferential, in the way that dream images are freed of any strict association with referents in the objective world. As such, it is a visionary glimpse of "la vraie vie" ["true life"],[16] which is "absente" (R 224) in the everyday reality of familiar occurrences and routines.

Plunging into the poem is like "passing through" a Blakean vortex or "descending" into Poe's maelstrom. Images of conical or spiro-helical turbulence abound in a context of clashing cross currents and polar antinomies. In order to reveal the symbol, I propose to disengage the essential rectilinear and angular structure of intersection, the alignment of explicit or implicit antitheses from which these conflictual currents originate and the synecdoches of a pervasive vorticity by means of which the disparate fragments and contraries are whirled into a synthetic, visionary whole.

MYSTIQUE

Sur la pente du talus les anges tournent leurs robes de laine dans les herbages d'acier et d'émeraude.

Des prés de flammes bondissent jusqu'au sommet du mamelon. A gauche le terreau de l'arête est piétiné par tous les homicides et toutes les batailles, et tous les bruits désastreux filent leur courbe. Derrière l'arête de droite la ligne des orients, des progrès.

> Et tandis que la bande en haut du tableau est formée de la rumeur tournante et bondissante des conques des mers et des nuits humaines,
>
> La douceur fleurie des étoiles et du ciel et du reste descend en face du talus, comme un panier, —contre notre face, et fait l'abîme fleurant et bleu là-dessous.
>
> On the slope of the knoll angels turn their woolen robes in pastures of steel and emerald.
>
> Meadows of flame leap up to the summit of the hill. On the left, the earth of the ridge is trampled by all the homicides and all the battles, and all the disastrous noises spin their curve. Behind the right-hand ridge, the line of orients, of progress.
>
> And while the band above the tableau is composed of the whirling and leaping hum of conch shells and human nights,
>
> The flowering sweetness of the stars and of the sky and of all the rest descends opposite the knoll, like a basket, —against our face, and makes the abyss fragrant and blue below. (R 283)

Structurally, an essential paradigm of the poem is the form "V" and its inversion "Λ" or upper-case Greek lambda, both of which depict the angular juncture of contrary trajectories. It is the latter structure that we find abundantly in the poem, represented also by implied pyramidal or conical forms. Sloping hillsides embody it in the images of "la pente du talus" and the "sommet du mamelon." The allusion to the female breast in addition to the geographical meaning of *mamelon* reiterates the conical shape. In the second alinea,[17] "le terreau de l'arête" repeats still again the mountain-hill image, and the word *arête* incorporates quite remarkably the paradigm of the angle of intersection both on the semantic level of the signified (i.e., the "rib," "ridge," or "bridge" of the intersecting surfaces) and on the purely visual level of the signifier (the lambda-shaped circumflex).[18] Finally, it is not unreasonable to suppose the "robes de laine" of the angels flaring out conically as they "tournent," completing the cluster of lambda-shaped synecdoches that predominate among the opening images of the poem.

Just as the angle of these figures seems to point upward, always emphasizing the summit, the image of "l'abîme fleurant et bleu," in the last section, points downward ("là-dessous") and occurs at the very bottom of the poem as a mirror image of the opening figures. Between these extremes, we discover the ambiguously orientated "conques des mers," which, although part of "la bande en haut du tableau," occur in one of the middle alineas of the poem.

The structural dichotomy of "Λ" and "V" is reflected on the thematic level by certain polar dualisms. First, there is height versus depth, as the previously mentioned allusions to *le sommet* and *l'abîme* indicate. There is also an opposition between left and right ridges ("A gauche le terreau de l'arête" and "l'arête de droite"). Associated with these directional loci are contrary cultures and values. The right side represents the east, "la ligne des orients, des progrès." Here, the straight, linear, progressive trajectory possesses positive connotations compared with the negative left side, which is "piétiné par tous les homicides et toutes les batailles." In this sense, the left (and, by inference, the west) seems also to oppose the destructive "past" of occidental history to the "future" promise of oriental culture toward the right. The statement that the left/west is where "tous les bruits désastrueux filent leur courbe" not only emphasizes the "sinister" qualities of that domain, but also places the sinuous "courbe" in opposition to the "ligne droite" of the east.

There can be no doubt that Rimbaud has portrayed both structurally and thematically the intersecting vectors of contrary forces, as embodied by the paradigmatic "V" and "Λ." But he has also supplied the synthetic whirl that results from the clash of these conflicting streams. Again, it is the parts that are immanent but that, when taken together, add up to the vortical whole.

Dynamically, the vortex is implied by various synecdoches of "circulation" and "spirality." The opening image of "angels turning their woolen robes" depicts a circular movement, which, if we imagine the skirts of the robes gathered at the waist flaring out like those of dervishes, mirrors the whirling conical form of the vortex. In the second paragraph, "les bruits désastrueux filent leur courbe," implying, once again, circular motion. But the key image of the poem in terms of vorticity is "la rumeur tournante et bondissante des conques des mers." The sea conches are conical objects but also, since they are turbinate shells, spiro-helical, and the spiro-helical pattern results from a motion which is, as we have here, both circular (*tournante*) and leaping (*bondissante*). The shells themselves, moreover, are frozen, calcified whirls, whose movement is conveyed aurally by "la rumeur tournante"[19] that is heard when such a shell is held up to the ear.

It seems clear when the parts are added up—the polar antinomies, the intersecting trajectories, the V- and lambda-shaped as well as conical forms, the circular and spiralic dynamics—that the *tourbillon* is implicitly the poem's central signified, which the poet has hinted at metonymically without actually providing the signifier ("Hortense found!"). The symbol also occurs in the context of juxtaposed natural (earthly) and supernatural elements, denoting a mystical cosmic breakthrough, as in Blake. Or perhaps, as Anne Freadman insightfully demonstrates, Rimbaud enacts an "apocalypse," involving both "cataclysm" and "regeneration," whereby the "original universe" is destroyed and replaced by a burgeoning "poetic universe"[20] (in which case the creative-destructive implications of the symbol are underscored). Actually, both readings are compatible and even complementary, since in order to achieve the creative, visionary penetration through to essences, one must first destroy the illusions and conventions of the immanent phenomenal world. As much as and perhaps more than any of Rimbaud's poems, "Mystique" testifies to the poet's attempt at such *voyance,* and it is the cosmic, turbulent vortical metaphor, I would argue, that he selects to symbolize it.

Over all, the Rimbaldian vortex is a metamorphosing entity, both patently and latently present, that develops across the young poet's career, becoming increasingly hermetic and complex. The important creative-destructive connotation is first only implied in "Le Bateau ivre," and only its destructive potential is evoked in the nihilistic Commune poem, "Qu'est-ce pour nous, mon coeur." Evolving simultaneously are associations with the vertiginous disorientation of the whirling dance and with the revealing distortions of dreams or drug-induced hallucinations. These are often the effects, no doubt, of the poet's calculated ("raisonné") "*dérèglement* de *tous les sens,*"[21] entailing a "derangement" or "disordering" not only of "all the senses" but also—as *sens* equivocally suggests—of "all meanings."[22]

In *Illuminations,* all of these fractured symbolic meanings converge and are combined into more complex poetic fabrics and structures, and this very confusion of disparate elements into a vision of transcendent unity is itself symbolized by the *tourbillon,* which with its whirl focuses all multiplicities upon the point of synthesis at the vortical core. Thus, all things correspond and are one. Just as "I is an other" and interchangeable with that other, so sea is earth, earth sky, past and future present, up down, right left, east west. But to glimpse this simplicity, one must pass through appearances to "la vraie vie" of pure essences. In the turbulent stream of temporal contingencies and events, the eddy formed by colliding currents, spiraling below the surface toward unknown depths, produces just such a breakthrough. What more vivid image, then, for Rimbaud's illuminating vision than these dazzling "vortices of light."

Whirling Toward the Void at Dead Center: Symbolic Turbulence in Mallarmé's *Un Coup de dés*

10

The symbolist concept of reality, based on the theory of correspondence, posits, in effect, a kind of Kantian dual structure, involving an ideal noumenal world beyond the imperfect, spleen-ridden world of phenomena. If for seers such as Rimbaud, the poet's mission is to penetrate through appearances to essences, in Mallarmé access to the "Idea" seems ineluctably futile in the world of phenomenal becoming, where chance prevails and cannot be "abolished" except through the fixity of death, humanity's only true "eternal" state, or perhaps through the imperfect eternality of art, which suspends chance and fixes a "piece" of being, for example, a thought, but as such is nonetheless an "absence" of the thing/event itself and at best an incomplete, approximate rendering of being. I must agree with Fowlie (echoing Thibaudet) that the title of the poet's remarkable last published work, *Un Coup de dés jamais n'abolira le hasard* [*A Dice Throw Never Will Abolish Chance*], the main focus of this inquiry into his recondite "univers imaginaire," "announces a failure,"[1] even if the poem itself seems ironically to stand as a paradoxical refutation of the very *échec* it depicts.

The failure takes the form, in the principal event of the poem, of a shipwreck. The ship's captain, as Gardner Davies interprets the dénouement,[2] bobs up briefly in the churning surf, dice clutched in his hand, and ponders uncertainly whether or not to throw them before he is himself sucked below the surface in the wake of the sinking vessel's whirling *tourbillon*. In such a way, the vortex structure functions as a central dynamic symbol, not only for the shipwreck episode but, as I intend to demonstrate, for the cosmic vision of the poem as a whole.

Before proceeding with an analysis of the symbol, a word must be said about the kinds of semantic and syntactic difficulties immediately apparent in Mallarmé's extraordinary text. Unlike the referential problem posed by Rimbaud, discussed in the previous chapter, the "syntaxer"[3] Mallarmé, especially in *Un Coup de dés,* conducts a daring linguistic experiment aimed at liberating thought from the tyranny of grammar by exploding the sentence into fragments that are then free to enter ambiguously into syntactic relationships with other such fragments dispersed strategically and meticulously across the page. In this respect, the poem is, in large measure, an epistemological statement that depicts the complex, fragmented nature of thought, visibly representing, by means of the ubiquitous "blancs," the absences and omissions that vitiate any complete expression of a thought and that disrupt the continuity that human reason seeks to impose upon the mind.[4] What Mallarmé is graphically illustrating on the page is, to use the geometrical analogy he himself has chosen, the "multi-dimensional" nature of thinking—the different size types add depth to the usual two-dimensional rendering of words on a page-plane—as well as the nonlinear dispersion of thought-fragments in the mind that, when expressed, are reduced, simplified, distorted, in order to conform to the linear, syntagmatic exigencies of grammar.

Thus, while Davies convincingly demonstrates that the lines can be read left to right, top to bottom, in his linear reconstruction of the syntax[5]—and this is indeed helpful—he also defeats Mallarmé's purpose of allowing for the kinds of free associations that occur in the mind as a thought takes shape. For example, the poet's isolation of principal subjects, such as "LE MAÎTRE" ["the master"] (4),[6] "*plume solitaire éperdue*" ["solitary bewildered feather"] (7) and "RIEN" ["nothing"] (10), allows them to be associated with any number of discrete thought-fragments, not just with the next one "in line" on the page. In the case of "LE MAÎTRE," in fact, the eye is drawn more readily to the important word "surgi" ["risen"] below than to "hors d'anciens calculs" ["outside old reckonings"], which although next in line, is more remotely placed across the double page.

The equivocacy in *Un Coup de dés,* then, arises more from the fragmentation of syntax than from the dubious possible "reality" of the referents, but the obstacles for the critic are nonetheless formidable. The highly compressed, economical, laconic, and at times mute nature of Mallarmé's *poétique,* with its preponderance of ellipsis and synecdoche, would seem to predispose it to a "paradigmatic" approach aimed at assembling and relating parts, which, when taken together, add up to a whole that may only be to a large extent implied. Accordingly, as before with Rimbaud, the vortex paradigm must first be established throughout the poem by disengaging its dialectical properties, e.g., opposing currents, intersection, concentric circularity, synthesis, vorticity. Only then can its significance in Mallarmé's work be analyzed and interpreted.

The polar structuring of the poem that creates the conditions for clashing currents and resulting vorticity is manifest in the most basic elements of its format, which the poet exploits in ways that still seem radical, a near century of poetic invention since then notwithstanding. That Mallarmé worked out early drafts on quadrille graph paper indicates the importance he attached to the horizontal and vertical dimensional poles, as well as to the third-dimensional axis of depth, created, as mentioned previously, by the varying type sizes, with the words in larger characters seeming closer than the more distant smaller ones. This effect of "perspective" is particularly apparent on page 9, where, in addition to several competing type sizes, a striking opposition between roman and italicized type is emphasized. In all, these spatial and typographical contrasts are pervaded by the poem's most fundamental dualism of black words (matter) arrayed on the white page (space), which is, in turn, reflected by the central dice-symbol's pattern of presumably black dots on white cubes.[7]

On the semantic and thematic level of the thought-fragments themselves, there are many oppositions. The very word occurs in the phrase "en opposition au ciel" ["in opposition to the sky"] of page 7, the so-called "Hamlet" page, where the "ideograph"[8] depicts—many critics seem to agree—the contrasting features of the Danish prince's headgear: a black "toque" sporting a wispy white "plume." The phrase may allude to the antagonism between passive fatalism ("a special providence") and the spur to action, which paralyzes Hamlet (note the word "immobilise" on this same page), but it also evokes the polar infinities of "sky" (Mallarmé's transcendent ideal) and opposing "sea" (where the disastrous shipwreck "takes place" amidst turbulent waves of immanent time-becoming).

Another important opposition exists in the equivocal character of "the as-yet-neutral word *JAMAIS*" (CM 24) of page 2, which resonates the positive connotations of the ideal "CIRCONSTANCES / ÉTERNELLES" announced on the same page and extends them suspensefully over two pages, only to be negated abruptly at the bottom of page 5. This "ever/never" dualism summarizes the principal existential dilemma of the "pensée" ["thought"] that the poem portrays. Its inherent ambivalence is reflected in several other ambiguous dualisms (variously recognized by critics): for example, "[la] voile" ["sail"] (3) and "le voile" ["veil"] (5), "la vague" ["wave"] (5) and "[le] vague" ["vagueness"] (10), the contrasts between light and dark in the fragment *scintille / puis ombrage* ["sparkles then overshadows"] (8), between beginning and end in the queries "COMMENCÂT-IL ET CESSÂT-IL" ["were it to begin and were it to end"] (9), between alternate sides in "penché de l'un ou l'autre bord" ["leaning to one or the other side"] (3), and in many other details. One of the most significant is the turbulent opposition between "hilarité" ["hilarity"] and "horreur" ["horror"] on the central page (6), which epitomizes at the work's core the positive and negative emotional extremes—laughter (cf. *rire* [8]), delight, comedy versus fright, misery, tragedy—that being, becoming through time, necessarily experiences.[9]

And this experience entails a dynamic alternation between poles, which is represented, on the poem's structural level, by oscillating wave patterns. Cohn in his exegesis has proposed various values for the relative positions of words on the page, including a positive (crest) and negative (trough) polarity (18).[10] There is no doubt that the title phrase traces a wave whose amplitude spans the vertical extremes of the page, although its frequency is just short of one full oscillation:

UN COUP DE DÉS

JAMAIS LE HASARD

N'ABOLIRA

That the movement from crest through trough ends at the neutral midposition without completing the cycle seems to me to illustrate two of the poem's central themes. The first has to do with the problem of "ending" that I shall take up later. The second insists upon the human plight of being trapped, like Hamlet, in a neutral position[11] between conflicting alternative choices, resulting in "bewilderment,"[12] hesitation, inaction. The crucial question for the "master"/captain/poet seems to be whether or not to throw the dice, and the various wave oscillations, including those down and up from page to page and sinuously back and forth on each page, the eyes tracing the zigzag of a meander, mimic the existential wavering that frustrates the poet's desire for any absolute determination. Hence, the isolated word "hésite" ["hesitates"] near midposition on page 4 announces the Hamletic quandary developed later on page 7 and invests the structural wave patterns with thematic significance. It is mirrored, on page 6, by the back and forth movements of the verbs "voltige" ["flits"] and "berce" ["rocks"], a "fluttering" and "rhythmic movement" Davies attributes to the feather floating above the abyss (124). But by far the most important echo of the hesitation theme is the verb "chancellera" ["will totter"] (5) describing the drowning mariner's dilemma as he wavers bewildered between the ultimate efficacy or futility of the dice throw, teetering physically on the brink of death. Occurring in the context of "s'affalera" ["will fall"] and "N'ABOLIRA," also in the future tense, the virtual nature of the act, should it ever take place, is emphasized, since it possesses, as yet, only the potentiality of thought.

In many ways, the alternation between extremes of the poem's central doubt recalls the to-and-fro movement that in Poe's tales precedes the moment of conjunction and crisis, symbolized by a descent into a maelstrom. Eventually, the wavering between poles yields to an encounter, a confluence of currents. In Mallarmé's poem such an intersection is imagined and represented, on the ideographic level, by the convergence of two word strands on page 8, in which Cohn sees "a Y-shaped figure . . . , tail of a diving mermaid" (12). Indeed, the word "*bifurquées*" ["forked"] occurs on the page with reference to the "scales" ("*d'impatientes squames ultimes*") of a "*sirène*."

On page 9, the two terms of a typical French *phrase hypothétique*, generated separately on the left side, converge in a complex dispersal of type faces and type sizes clustered on the right about the riveting word "LE HASARD," with which the title phrase finally culminates after a three-page hiatus. Besides reiterating the hypothetical (virtual) nature of the dice throw, the imperfect and conditional verbs are both forms of the verb "to be":

> *SI* . . . *C'ÉTAIT* . . . *LE NOMBRE* . . . *CE SERAIT* . . . LE HASARD
>
> [If it were the number it would be chance][13]

The confusion, here, of being and number/chance might seem faintly to echo Baudelaire's complaint, at the end of "Le Gouffre," "—Ah! ne jamais sortir des Nombres et des Êtres!" [Ah! never to be free of Numbers and Beings!], but as one critic implies,[14] Baudelaire yearns to escape insentiently into the very nothingness that Mallarmé hopes, through the permanence and certainty of number, to defeat and transcend.

On the thematic level, the clash of cross currents that symbolize in sea imagery the Hamletic internal conflict of the drowning mariner's doubt in the face of death—later called "la mémorable crise" ["the memorable crisis"] (10)—is described as "induction" "vers cette conjonction suprême avec la probabilité" ["toward this supreme conjunction with probability"] (5). Again, the virtual ("probable") nature of the conflict is stressed. Previously, on page 3, the inevitable conjunction-crisis was subtly prefigured in a nautical term, *l'envergure.* The "span" of the sail is an image of connection and cohesion, containing the root word, *vergue,* the yard from which the sail is suspended. On ancient Greek ships, the central part of the *vergue* was called a "symbol," because it was the point of intersection of the yard arms and of the mast's rigging.[15]

One last image of this "symbolic" conjunction, which embodies the essentially rectilinear structure of intersection (like the cross formed by mast and yard together) but denotes, at the same time, a synthesizing of contraries by means of circular motion, is the dice throw itself. The opposition of white cubes and black dots has already been mentioned, but we have here, as well, the act of "rolling" the (square) cubes as if they were (circular) spheres. A "coup de dés" seems to be, in this respect, an attempt symbolically to reconcile (earthly) rectilinearity and (heavenly) curvilinearity. The awkward tumbling of the scattered cubes—"le heurt successif" ["the successive shock"] (11)—is an apt symbol of the "hard knocks" that being experiences while trapped in earthly time-becoming. The ultimate resolution of such "hasard," however, is signified by the "heavenly" round dots which, when the dice come to rest, fix eternally the Idea, i.e., "l'unique Nombre qui ne peut pas être un autre" ["the unique Number which cannot be another"] (4).

The theme of synthesis, to complete the dialectics of the vortical paradigm, is a recurrent motif throughout the poem, conveyed on page 3 by the verb "résume" ["resumes"], by the phrase "en reployer la division" ["refold its division"] (4), by the allusion on the next page to "Fiançailles" ["betrothal"] (a literal marriage of opposites), and by one of the central notions of the last page, "fusionne avec au delà" ["fuses with beyond"], which summarizes pithily the transcendent synthesis that the dice throw is intended to accomplish. And Mallarmé symbolizes this impulse toward the absolute beyond by means of pervasive circular movements and structures.

Nearly all the important studies of the poem analyze, to some extent, its circularity. Ending as it begins with the phrase, "un coup de dés," an overall circular structure is obviously implied. Davies argues that "*Le Coup de Dés* in its entirety takes the form of a gigantic chiasmus" (79),[16] illustrating it as follows:

UN COUP DE DÉS–*COMME SI*–*COMME SI*–UN COUP DE DÉS[17]

The mirror imagery of this fundamental structuring denotes, in effect, an inner and an outer circle. The former occurs on the innermost page (6), and its ideograph designs, appropriately, a circular disk viewed roughly side-on. Five double pages then fan out symmetrically on each side, like ripples on a pond when a stone has struck the surface, culminating at the extremes (pages 1 and 11) with the dice-throw phrase.

Another chiasmus is the curious phrase of page 5: "la mer par l'aieul tentant ou l'aieul contre la mer" ["the sea by the ancestor trying or the ancestor against the sea"]. Here, however, an opposition is implied by the prepositions *par* and *contre,* evoking, it would seem, the dice thrower's ambivalent attitude, in both *Un Coup de dés* and *Igitur,* toward his ancestors. The fusion of the sea and ancestral[18] themes in this circular phrase seems to compare the generational cycles of his family line to rolling waves, cresting and dissolving insignificantly in a vast sea of space-time.

Two of the poem's "orphic"[19] pronouncements display the circular logic of paradox and seem, consequently, to admit myriad interpretations. The most prominent of these is the title phrase, "UN COUP DE DÉS JAMAIS N'ABOLIRA LE HASARD." The circularity derives from the fact that a dice throw, which invokes (puts into play) the operations of chance, is proposed as a means for destroying (surpassing, getting free of) chance. The title phrase is thus the poet's negative response to a question that asks, in effect, "can chance be used to abolish chance?"[20] Similarly, the tangential pronouncement, "RIEN N'AURA EU LIEU QUE LE LIEU" ["nothing will have taken place but the place"], not only posits the absurd coming into being of an a priori condition of being (the space of "place"), but also evinces the circular logic of a tautology by stating, in effect, that space will have occupied space.

Perhaps the most subtle, if nonetheless symbolically charged, circular image, is the allusion right at the outset to "CIRCONSTANCES / ÉTERNELLES." This is indeed a succinct evocation of the kindred concepts of circularity and transcendence. "Circum-stance," recalling again Michel Serres's analysis, is just the right word to express Mallarmé's conception of existence outside of space and time (which is only hypothetical, as the uncertain "MÊME" ["even"] implies). The perfect and immutable symmetry and cloture of the circle ("circon-"), with neither beginning nor end, is a standard symbol of the eternal. Coupled with the anthropocentric, perspectivist connotations of "-stance," it very aptly expresses the poet's desire to attain a concomitantly human and transcendent ideal. But it also connects syntactically and paradigmatically (through the implied circular tracing of a circumference) with the image, on the same page, "DU FOND D'UN NAUFRAGE" ["from the depths of a shipwreck"], bringing us finally to the symbol toward which all of the dialectical elements we have heretofore been examining converge.

The sinking ship, as we have seen, creates a whirling eddy in its wake toward which the captain/master, surfacing for a moment amidst sundry flotsam, is inexorably drawn. Lest there be any doubt that the shipwreck involves vorticity, notwithstanding the turbulent conditions just established, the poet develops a cluster of vortex images: the virtual abyss in "SOIT / que / L'Abîme" ["whether the abyss"] of page 3, intensified by the double occurrence on the same page of "profondeur" ["depth"]; the repetition of "gouffre" ["gulf" or "abyss"], first as the terminal image of the "HASARD" page (9) and then closer to the vortical core of the work, literally "autour du gouffre" ["around the abyss"], on the central page (6); finally, at nearly the precise center of the edge-on disk that the "*COMME SI—COMME SI*" circular ideograph designs, the dynamic "tourbillon" (about which the entire poem turns) emerging dialectically at the pivotal juncture of contrary currents, symbolized by the opposite, though equal, qualifiers, "d'hilarité et d'horreur."[21]

However, the dissolving vortical streamlines that the ideograph suggests—the italics heighten the shimmering, ephemeral effect—represent not the drowning itself in which "la mémorable crise" culminates, but either the victim's view as he approaches the whirlpool, in which case it is virtual, or, more ostensibly, the aftermath of the spiro-helical descent, which the reader alone beholds, its insignificance underscored, for the ocean turbulence will soon erase all traces of the event, that is, "COMME *SI* . . . RIEN N'AURA EU LIEU QUE LE LIEU" ["as if nothing will have taken place but the place"].

That "a descent into the maelstrom" is implied, whether or not it does actually occur, is indicated by many factors, and in this respect, Mallarmé's text resembles Lucretius's cosmic epic, *De Rerum Natura.* One critic was struck by this comparison at a time when he was reading simultaneously

Mallarmé and Serres's work on Lucretius.[22] Turbulent swarms of motes ("mots") do seem to form and dissipate in the white void-space, as the eye roves from page to page of Mallarmé's own epic portrayal of "the nature of things."[23] And the movement is always downward and sideways, "sous une inclinaison" ["under an incline"] (3), as if to illustrate the oblique Lucretian "swerve" of atoms—"selon telle obliquité par telle déclivité" ["according to such obliquity by such declivity"] (11)[24]—that arises spontaneously by "chance" to form, however fleetingly, aggregates, systems, worlds. No gods, no afterlife (and thus neither hope[25] nor, for Lucretius at least, dread); only the inevitable, if as yet virtual, fall—"s'affalera" (5)—after teetering precariously—"chancellera" (5)—on the brink. Like the "bewildered solitary feather," fluttering, suspended vertiginously above the whirl, all swarms and systems (every *turbo*) must ultimately "decline," perish, dissipate, be swallowed up by the "sinister" depths, return to the originary chaos (*turba*) whence all things spring, or in Mallarmé's own concise word-image:

Choit
la plume
rythmique suspens du sinistre
s'ensevelir
aux écumes originelles

> [Falls the feather rhythmic suspense of the sinister to be buried in the primal foam] (9)

It seems clear, then, that the "descent into the maelstrom" portrayed (imagined) in *Un Coup de dés* represents for Mallarmé the inevitable decline toward death-extinction to which all beings, becoming through time, are doomed. Death is, in fact, the unnamed central theme—absent because it is an absence—symbolized by the poem's central *tourbillon*, which, in nature too, turns about a vacuum-void. Still, even if the word itself does not appear, allusions to death abound: the two mentions of the "naufrage" (2, 4), the "cadavre" floating among the word-fragments (flotsam) of page 4, the references on the next page to "la disparition" [the disappearance] and "le fantôme d'un geste" ["the ghost of a gesture"], the profusion of negatives as epitomized by the word "RIEN," set off in initial position on page 10, and finally, on the same page, "l'absence" that *is* death, symbolized by the vacant seascape ("en quoi toute réalité se dissout" ["in which all reality dissolves"]) after the whirling has subsided, and nothing is left but the sheer emptiness of "place."

But the poem really focuses, as I have suggested, on the decline toward death, on dying, since, as Samuel Beckett has dramatized more recently, the "end" of the "game" never takes place (Godot never arrives); there is only ending, waiting. Analogously, one never really knows death when it occurs, because one is dead (i.e., death remains ever virtual until too late to matter). It is rather the anticipation of death—dying—over which we agonize. Mallarmé vividly portrays this progression toward death-absence in the rolling plunge of present participles—admired by Gide—that conveys the poem "toward" its conclusion (although, again, this never quite takes place, due to the overall circular structure):

veillant
doutant
roulant
brillant et méditant
avant de s'arrêter
à quelque point dernier qui le sacre

> [watching doubting rolling shining and meditating before stopping at some final point which consecrates it] (11)

A life lived is like the casting of dice: the outcome is uncertain, subject to peripeteia and contingency, as long as the operations of chance through time-becoming are in play. Only when the tumbling dice come to rest, when death occurs in the depths of the maelstrom, is "the unique Number that cannot be another," the life's "essence," to use a Sartrean term, revealed. This essence, as immutable as a "constellation," is nonetheless a legacy—"legs" (5)—and, as such, escapes, surpasses the being, now dead, whom it describes. Consequently, Mallarmé's transcendent aspiration ends in personal failure—an individual's only eternal state is the insentience of death itself—even if, by an act of art, a constellation survives as a legacy for posterity.

But the drama of the poem centers about the penultimate moment before the final dice throw. The full wave period of the title phrase, as we have seen, has not yet returned to the crest position, completing the cycle. More than death itself, which remains virtual (as the myriad hypothetical terms, future and conditional tenses, and subjunctives attest), it is rather the "thought" of extinction that bewilders the master/captain/poet, who, like Hamlet, vacillates between equally disagreeable alternatives, since in his epicurean view, there is nothing to hope for beyond death's finality. In *Igitur,* this frustrating uncertainty had driven Mallarmé's protagonist, again like Hamlet, to thoughts of suicide, imagined as a Poesque descending of a "spiral" stair—"la spirale vertigineuse conséquente" ["the resulting dizzy spiral"] (M 437)[26]—to the family crypt below. The act seems to be accomplished in the last scene of the work, captioned "Il se couche au tombeau" ["He lies down in the tomb"], by ingestion of poison—"la goutte de néant" ["the drop of nothingness"]—from a "fiole" (443). The poet, playing anagrammatically on the word ("La fiole vide, folie . . . ") ["the empty phial, madness"], seems to relate the suicide to the madness announced in the work's subtitle (*Igitur ou la Folie d'Elbehnon*). In effect, the destructive Hamletic quandary must issue, as it does in Shakespeare's play, in the madness of suicide (Ophelia's case) or in a surrender to fate (providence) that ushers events inexorably to their tragic conclusion.

If *Igitur* explores the first solution, *Un Coup de dés* illustrates the inevitability of the second. Although madness is a menacing contingency, occurring in the depths of a wave-trough (5), there is no phial, no meditation upon suicide in *Un Coup de dés,* precisely because absorption into the vortex is imminent. In this respect, the master's indecision concerning the dice throw is moot. The dice will, must be cast at the moment of death, when becoming takes on the eternal essence of being, when the thought is formulated, fixed, frozen in the poem's constellations of words, a legacy of the struggle that only the descent into death's vortex could resolve. *Un Coup de dés* then, unlike *Igitur,* is tragic. The latter involved a free choice; its outcome might have been avoided. The former, on the other hand, portrays the unrelenting pull of the expanding streamlines toward the void at dead center: a dire fate, like Hamlet's, that cannot be resisted, as master, poet, and by extension, every human being experiences the fatal "naufrage" that, like the Lucretian *clinamen,* must "necessarily" occur sooner or later in the course of contingent events, engendering in its wake an inescapable "descent into the maelstrom" that dooms all, willing or not, to eternal nonentity. Accordingly, although death delivers ("émet") the final "coup de dés" for each life, checking, defeating, annihilating "le hasard" and all possibilities of change through time for that individual, it does not, cannot, abolish chance altogether; it "never will."

When the three authors we have just studied are compared on the basis of a symbol common to them all, certain metaphysical and aesthetic principles of the symbolist episteme emerge. Reality is, above all, double, and consciousness, discovering itself trapped by matter in a chaos of multiplicity, yearns for the serenity and simplicity of the pure idea. Only the poet, a privileged seer, is able to effect such a transcendence through an act of art, predicated upon the alchemy of language, which transmutes base experiential matter into a finer, rarer essence. Often, the transformation is symbolized by a penetrating whirl, denoting a dynamic, radical metamorphosis from one state or dimension into another, although the significance of the symbol varies from author to author and usually betrays characteristics peculiar to his own conception of turbulence and the transcendental ideal.

For Poe, who is an obvious seminal influence upon the others, the vortex symbol resonates many meanings on various levels. It is, first of all, a powerful natural force that indifferently (and even malevolently) draws hapless victims slowly but surely—and thus with agonizing suspense—to their doom. It is the pit and the pendulum, the former because it is an environing hole into which one "descends," the latter because the descent, the approaching destruction, proceeds "by degrees." It mirrors, in this sense, Dante's abysmal hell, swallowing up the wicked, threatening all who come within its influence, although a just, resourceful person might escape its clutches if he keeps his wits. And yet the sheer power of its turbulent fury is a source of awe, holding Poe's narrators enthralled, piquing their curiosity, inciting within them the "imp of the perverse" that impels them willingly to jump in. As such, it is a surrender to the irrational, the manifestation of a death instinct, a passage into pure imagination that holds out the thrilling prospect of "discovering" something "novel," a profound cosmic secret, the knowledge of which seems well worth the dizzy plunge toward death.

Rimbaud's turbulent images denote similarly an epiphanal breakthrough and the discovery of something new, but not so much via death as through mystical hallucinations induced by a calculated debauchery and the intoxication of poetry. In *Illuminations,* the vortex is a devastating force only to the degree that it signifies an assault upon bourgeois conventions of a repressive and corrupt culture, expressed as violence against the language that enshrines and perpetuates those conventions. By means of the whirl, signifiers are wrenched from signifieds, signs from referents, as the poet-seer vertiginously penetrates the illusions of an artificial order in search of new paradigms and an essential unity. As in Baudelaire, correspondences are discovered beneath a surface reality of multiplicity and difference. But Rimbaud is not content just to record the experience of his illuminating vision. He hopes, by creating a language capable of conveying it in all its fullness, immediacy, and intensity, to effect new possibilities of experience and radically change the world.

Mallarmé shares Rimbaud's desire to access a realm of pure idea, but no such conscious transcendence is afforded by subsumption into the vortex, even if human being does achieve, in death, the imperturbable ideal it unsuccessfully sought in life. The "descent into the maelstrom" is thus for him a lethal, calamitous plunge, as in Poe, but no "imp of the perverse" nor any keen curiosity about the unknown conveys the master/poet willingly to the brink. Rather, he is tossed about helplessly, haphazardly in a vast sea of contingent time-space, while the vortical core's vacuum-void draws him inexorably toward ultimate extinction. And even if the poet is able to consign to the ages a legacy of beautiful works, they are only as immortal as the ages themselves and as perfect as the flawed, frustrated genius who fashioned them. Mallarmé's *tourbillon* is thus like Baudelaire's *gouffre,* a negating absence at the center of all experience—"action, desire, dream, / Word"—about which life itself whirls in ever-narrowing concentric circles through time, proceeding (descending, declining) centripetally, vertiginously by degrees toward a fatal vanishing point of no return. Unlike Baudelaire, though, who envisages in death's abyss a Poesque voyage "into the depths of the Unknown to discover the *new!*"[27] Mallarmé discerns no such continuation of consciousness or of the poetic adventure, only a manifestation through art of the self-become-idea—"Tel qu'en Lui-même enfin l'éternité le change"[28]—as fixed and final, enduring and inert, calm and cold as a tombstone.

Conclusion

I end with Mallarmé in *fin de siècle* France, not because *Un Coup de dés* represents an apotheosis of the *tourbillon*, although one would be hard-pressed to find a more brilliant poetic evocation of it or a more powerful image upon which to conclude. Symbolic turbulence has continued to thrive and develop in the twentieth century. "Vorticism," an offshoot of Cubism, adopted the vortex as the emblem for a new aesthetic reflecting the radical creative and destructive changes of the period leading up to and culminating in the First World War. The poetry of W. B. Yeats is another salient source of spiro-vortical symbolism in modern times, and the "gyre" is for him a symbol with profound historical and metaphysical significance. To have pursued these and other leads, however, would have required many additional chapters on subjects that, fortunately, have already elicited considerable scholarship and research. I have preferred to discover the symbolism where it has not already been noticed or to suggest new interpretations when its meaning has not been convincingly adduced. I have also sought to demonstrate the ubiquity and complexity of the vortical paradigm, as well as its capacity, as a symbol, to encompass a multiplicity of dialectically emerging and receding, diverse and even contradictory significations.

This is not the place to attempt a detailed review of the entire catalogue of meanings that have accrued in the course of my study, since, as they say, "once is enough." But I do think one last evocation of the symbol, in retrospect, will provide an appropriate final synthesis and even bring to light new points of comparison and contrast.

Aesthetic turbulence first emerges in ancient "whorls" and "whirls" that often display a spiral, spiro-helical or vortical configuration. The asymmetry of these forms invests them with dialectical symbolic potential. If one considers only the fundamental structure common to them, that is, cycles whirling out from or in toward a point of origin or termination, a progressive polar reversibility is immediately apparent, which accounts for their ancient association with such concepts as birth and death, creation and destruction, and processes that proceed by degrees.

Curiously, two of the most significant spiro-vortical phenomena are sea-related, the turbinate seashell and the whirlpool, which exemplify the ambivalent nature of the symbol. Both, as we have seen, can represent birth, as the two versions of Aphrodite's emergence from, alternately, a shell or a whirlpool demonstrate. But the protective function of the *coquille* as a refuge against predators is quite the opposite of the maelstrom's annihilating effect upon things coming within its influence, the one symbolizing preservation, the other destruction.

In the earliest textual examples, it is the destructive connotation of the whirlpool (and whirlwind) that stands out and links the vortex to dangerous, powerful (natural and supernatural) forces. In the case of the "Boulak Papyrus," a disruptive human threat is signified by the whirlpool, and the tone is morally reprobative and xenophobic. The Bible and Homer's *Odyssey* personify the whirlwind and the whirlpool as powerful and potentially destructive superhuman forces. That the Old Testament deity, Yahweh, appears "in the whirlwind" is striking evidence of that symbol's association with a (con)fusion of the transcendent and the immanent, heaven and earth.

The formation of eddies in turbulent fluid flows invests the vortex with connotations of order emerging out of chaos. It is this aspect that interests the ancient Greek philosopher-poets we examined, who develop the concept of a constructive, cosmogonic "whirl" that, by means of a principle of organization, caused things to come to be the way they are. Each conception of the whirl betrays a particular vision of the force(s) at work in the cosmos. Plato's account is the most rationally ordered, and the pure and perfect forms that underlie the imperfect system of concentric hemispheres of his "Great Whorl" are, in effect, a culmination of the idealism implied in the earlier concepts of Heraclitus's "Logos," Empedocles' "Love and Strife," and Anaxagoras's "Mind." The atomist and Epicurean-Lucretian views, on the other hand, are more materialistic and rely rather on chance or the "necessity" of natural laws to explain "the nature of things." In Lucretius's cosmology, contingent swerves of atoms engender and destroy systems, states, worlds through vortex action in a violently turbulent universe of falling and colliding motes. Thus, although the underlying causes vary, the vortex, as a pattern and/or process of organization, is, for all these philosophers, a universal structuring principle.

So it is, too, for Dante in his conception of the three dimensions of the afterlife: heaven, purgatory, and hell. The emphasis, however, is no longer on the dynamic cosmogonic role of the whirl, but rather on the levels and degrees inherent in the vortical structure itself (apt for the depiction of Dante's hierarchies of rewards and punishments), as well as on dialectical and reversible characteristics of the vortex. Through transformations and inversions of the figure and its properties, the poet is able to portray the descent into death of hell's concave pit, the cleansing ascent toward life of purgatory's enantiomorphically convex cone-mountain and the attainment of eternal being and bliss in paradise's harmoniously gyrating system of concentric heavenly circles: all incorporating variations of the same essential vortical form. Moreover, the structural oppositions are portrayed in terms of the moral categories of good and evil that pervade the Christian ethos. Even directional indications evince moral values, although, in Dante, perspective is all-important in determining these values, since he places his symbols not only in both hemispheres of the earth (where the connotations of "left" and "right" are reversed) but also outside of time and space. Generally, witherwise motion denotes for him a negative course or the undoing of one, and sunwise motion a positive (beneficial, redeeming) trajectory.

For both Descartes and Blake, the moral typing is less complicated: left and right are, respectively, negative ("sinister") and positive ("righteous") directional vectors, but whereas Descartes's values are Catholic and thus compatible with Dante's, Blake's maverick theology produces symbols of quite a different sort, such as the association of the witherwise spiral or vortex with Jehovah, the Church, and scientific rationalism, demonstrating how radically he departed from tradition—both Catholic and Protestant—in his own "reformation" of the Christian faith.

Among these authors, Descartes is the closest in spirit to the Greek thinkers, especially in his vortex theory of planetary motions. Like them, his system of cosmic whirls proved scientifically fallacious, but like them, he did intuitively grasp the right structure, even if he applied it erroneously to the planetary rather than the then-unknown galactic level. The Pre-

Socratics and Descartes, one could say, were accurate symbolically, if not scientifically, and they were all, in this respect, cosmic poets.

Descartes's description of a transcendent creative "enthusiasm" associated with the experience of a vertiginous vortical spin also recalls prophetic subsumptions into the divine whirlwind in the Old Testament and echoes Dante's association of visionary transcendence with the gyrations of vortical motion. Blake avails himself repeatedly of the vortex symbol, as we have seen, to denote a visionary breakthrough from one zoa to another. In fact, it is the elaboration of this epiphanal vortex that sets the thinkers of the Christian ethos apart from the classical Greeks, whose *dinos* was intended more as a mechanical cosmogonic model, even if it was charged with symbolic multiple-sense. The epiphanal connotation also links the Christian ethos with the subsequent "symbolist" portrayal of the paradigm, although again, the symbolic associations are significantly transformed.

In Poe, the symbolic ambivalence of the vortex continues and is even underscored in his depiction of the maelstrom. In a way, it represents a not specifically religious hybrid of Dante's hell and heaven. On the one hand, it is a gaping abyss (into which one falls, descends, is drawn toward death) and a potential dispenser of a kind of natural justice (the case of the resourceful narrator's and his treacherous elder brother's deserved fates in the "Descent"); on the other hand, it is a passageway to the "beatific vision" of the transcendent, eternal "Unity" (Godhead) whereupon all phenomenal multiplicities converge. Hence, the maelstrom for Poe's narrators excites conflicting feelings of horror and awe, as they waver between a self-saving desire for life and an "imp of the perverse" piquing their curiosity about the "novel" beyond, even at the risk of death.

With Rimbaud and Mallarmé, the effects of turbulence are apparent on the level of language itself, as though a tornado had beset the complacent calm of a vegetating grammatical field, wreaking havoc—wrenching signs from their referential roots in Rimbaud and scattering syntactic debris pell-mell in Mallarmé. In *Illuminations,* this disruption of the sign system that represents the conventions of a bankrupt culture is an act of sabotage directed against those conventions by a linguistic terrorist bent on changing the world by changing the conception of the world. Such violence against the status quo is a necessary precondition for the revaluation of all values that can only be achieved by getting beyond appearances to essences through an act of revolutionary poetic imagination.

In this respect, the turbulence of Rimbaud's visionary adventure resembles the breakthrough into a new dimension of Blake's epiphanal vortex but also holds the promise of a "novel discovery," as in Poe.

The "descent into the maelstrom" in Mallarmé's *Un Coup de dés* is Poesque in the more negative way. Not only is it a threatening destructive force, it is the very specter of death itself, relentlessly drawing the shipwrecked mariner, the poet, humanity to ultimate extinction and nonentity. It recalls, in this sense, Lucretius's turbulent atomic swarms, ever declining, dissolving, dissipating in a chaotic universe abandoned to chance. For Mallarmé, the *tourbillon,* like the contingent dice throw, whirls furiously in the sinking ship's wake—the dice rolling, tumbling awkwardly until, slowing, they come to rest, inert, the unique eternal number fixed—while churning surf, settling, subsides, nothing surviving but the place, unless perhaps a legacy. If for Lucretius the message of death's finality is glad tidings, freeing him from worry about an afterlife, it is for Mallarmé a frustrating and ironic paradox—a limiting, inhibiting negation that checks all efforts of the living self to attain the pure idea until, dead (and thus too late), the self itself becomes its own idea.

When all of these examples of vortical turbulence are considered in a last retrospective glance, one observes the elaboration of a problematic of being on the physical level of a phenomenon and the metaphysical level of a form, which continues to fascinate physicist and metaphysician alike. In nature, the shifting states of order and chaos in a turbulent flow are still not fully enough understood to be predicted, and the astrophysicist, pondering the contradictions of a black hole, confronts in its vortical depths the horizon of the known. Here we have two cogent examples of the vortex mediating between extremes, demonstrating how readily it lends itself to adaptation as principle, process, metaphor. No wonder that, early on, it caught the attention of the philosopher, stimulated the imagination of the poet, as each sought to explain the way things are. And its evolution as a trope has engendered, as we have seen, a dynamics of emerging and receding meanings welling up from a seemingly inexhaustible reservoir of significance.

But there does seem to me to be one particularly salient feature of the symbol that stands out among the others, since it is common to them all, which is the one just mentioned, namely, its function as a dynamic principle of mediation. Halliburton touches on this when he notes, in his study of Poe, that the vortex is a "space-toward" and a "space-between." We always find it at the limit of things, at the point where one thing becomes another—life death, immanence transcendence, reality a dream—and vice versa. In a black hole, even space itself (and time and matter) whirl toward their negation, as all the givens of the real converge upon a vanishing point of infinity. As a locus of mediation—here I must assume these elements of the a priori to continue—the vortex is thus a threshold between here and there, before and after, and any object crossing over is utterly transformed. In this respect, it is a symbol par excellence of change, but more precisely of a dynamic process of change, since the transformation happens by degrees. Because the there is inaccessible to all but the one spiraling through, the end result of change is unclear, making the pit of the whirl an ambiguous space charged with mystery. Whether we shrink back in horror at the terrifying sight of it or lean forward out of curiosity about whatever lies beyond, the transformational vortex, conveying us to the very brink of the known and the unknown, dares us to "break on through."

Appendix: Vortices, Helices, Spirals, and Gyres

The four terms of this title may seem synonymous to many, and they are frequently used as if they were. To my knowledge, no comprehensive or comparative study of these structures exists, and those that deal with one or another of them individually tend to produce conflicting and even contradictory definitions. In order to bring consistency to my discussion of symbolic turbulence, which necessarily involves, to varying degrees, all four configurations, I propose to examine each in detail, defining it in relation to the others and pointing to specific examples of it in the cosmos. I say "cosmos" because these forms exist at all levels of the known universe, widely dispersed through the whole range of phenomena from the micro- to the macrocosmic, and inhering in both organic and inorganic systems and states.

The Spiral

If we define a spiral as "a continuous curve traced by a point moving round a fixed point in the same plane while steadily increasing (or diminishing) its distance from this,"[1] the "Archimedean" or "equable" (TG 752) spiral is one in which "the perimeters of the spiral remain parallel,"[2] as in the case of "the coil of the rope or the whorl of the young tomato plant." Moreover, "A spiral of Archimedes, if a straight line is drawn from its origin outward, intersects that line always at the same interval from one whorl to the next" (JS 36–37); and such a radius not only "will increase in *arithmetical* progression" from whorl to whorl, but "meets the curve (or its tangent) at an angle which changes slowly but continuously, and which tends towards a right angle as the whorls increase in number and become more and more nearly circular" (TG 752–53).

In the case of the "logarithmic" or "equiangular" spiral, "it is the angle that the perimeter makes to the radius that remains constant as the spiral expands" (JS 36), as exemplified by a "chambered shellfish" like the nautilus. D'Arcy Thompson further notes that, for such a spiral, "Each whorl which the radius vector intersects will be broader than its predecessor in a definite ratio; [and] the radius vector will increase in length in *geometrical* progression, as it sweeps through successive equal angles" (753). Thus, while the resulting curve is asymmetrical, it displays the "fundamental and 'intrinsic' property" of "continual similarity" by means of which it "grows in size *but does not change its shape*" (TG 757). In the case of a chambered shellfish,

> the shell retains its unchanging form in spite of its *asymmetrical* growth; it grows at one end only, and so does the horn. And this remarkable property of increasing by *terminal* growth, but nevertheless retaining unchanged the form of the entire figure, is characteristic of the equiangular spiral, and of no other mathematical form. It well deserves the name by which James Bernoulli was wont to call it, of *spira mirabilis*. (TG 758)

Hence, the logarithmic (equiangular) spiral, which "was first recognised by Descartes, and discussed in the year 1638 in his letters to Mersenne" (TG 753–54), is not only "one of the most magical curves in geometry," but is of central importance in biology as well, since it is "the spiral most characteristic of the forms of living things" or "the spiral of growth" (JS 37), and as Thompson puts it, "Somehow or other . . . the *time-element* always enters in" (752).

There are, of course, many forms of spirals, of which the aforesaid are the two most widely dispersed in nature. A comparison of the configurations of spiral galaxies, for example, would attest to this diversity. One of three general classes of galaxies—the other two are described as "elliptical" or "irregular"—the spiral type, of which fully a third are "barred spirals," range widely in size and shape, from those with massive centers and short arms to those with tightly compacted centers and long, trailing arms:

> In both normal and barred spirals we observe a gradual transition of morphological types. At one extreme, the nucleus is large and luminous, the arms are small and tightly coiled, and bright emission nebulae and supergiant stars are inconspicuous. At the other extreme are spirals in which the nuclei are small—almost lacking—and the arms are loosely wound, or even wide open. In these latter galaxies, there is a high degree of resolution of the arms into luminous stars, star clusters, and emission nebulae. Our galaxy and M31 are both intermediate between these two extremes. (AR 334)

The same sort of variety might be discerned in a survey of shellfish or of animal horns, which also frequently assume the spiral form. But all spirals share in common the basic structure of a center and periphery linked by convolutions of a coiled line segment.

It should be noted that the term "spiral," for the purposes of this study, refers to, essentially, a two-dimensional figure in which parallel convolutions unfold progressively along a plane.[3] This restriction of the sense is consistent with the OED's association of "spiral" with that "which is applied only to plane curves" (1285). Thus, the alignment of stars in the Milky Way across a spiro-circular plane, referred to by astronomers as the galaxy's "disk," is primarily spiralic—even though it is a question of three-dimensional objects distributed in space—since the spiral arms swirl out in a plane about the central nucleus. The same is true of the solar system, although, in this case, the spiralic movement is less apparent. Arrayed in concentric circles along the plane of the "ecliptic," the planets, particularly those nearer the center, are gradually spiraling centripetally toward the sun, just as the entire solar system, located in a spiral arm near the periphery, is slowly spiraling, at a rate of one rotation every quarter billion years, toward the galactic center.[4]

Another three-dimensional object that must be classed, in my taxonomy, with two-dimensional spirals is the rolled-up scroll or volute. Like the galactic spiral, it is essentially planar, since its extension into the third dimension in no way affects its structure other than thickening it, as a simple thought-experiment will demonstrate. Imagine four objects arrayed in the following sequence: a spiral drawn on a sheet of paper, a phonograph record, a coiled tape measure, and a rolled-up rug stood on end.

All four items evince the same convoluted configuration, and all four are three-dimensional, although the thickness of the drawn figure is nearly undetectable, while the thickness of the other objects is progressively more apparent. All of these are nonetheless simple spirals, because the expansion of the spiral along the center-become-axis adds nothing new to the shape besides bulk. Severed at any plane perpendicular to the axis, the cross section will yield the same spiral, as if one were to divide up a jelly-roll pastry. Thin slice, thick slice—it would amount simply to smaller or larger quantities of the same essential form. The fact is that two-dimensionality can exist only as an abstraction in a three-dimensional (or four-dimensional) reality, so a structure is spiralic only to the degree that the third dimension approaches or conforms to a plane, *relative* to the rest of the structure's distribution of parts.

With this important distinction understood, we can survey spiralic phenomena without concern for the apparent contradiction of citing three-dimensional objects as examples of a two-dimensional structuring principle. And one notes right away that any such survey reveals an uneven distribution of the form, with concentrations much greater in some areas than in others, despite its universality. A case in point is the entire phylum Mollusca of shelled, soft-tissue invertebrates that manifest a distinct preference for the (preponderately logarithmic) spiral configuration as a principle of growth (e.g., chambered shellfish like the nautilus, snails, and other mollusks on land and in the sea) or as a means of arranging limbs or appendages efficiently (the case of the octopus, "which in repose curves its tentacles in spiral form" [MM 49]). There are, too, the Foraminafera, a group of tiny one-celled, shelled sea creatures (Protozoa) displaying myriad variations of the spiral structure, to which D'Arcy Thompson devotes his entire twelfth chapter. Indeed, the high incidence of the spiral symbol among maritime peoples has been associated, by some theorists, with the abundance of the form in sea-related phenomena.

Other expressions of the spiral as a principle of growth are displayed in wood whorls surrounding knots in trees (like the whorls of human fingerprints)[5] and in leaf arrangement or phyllotaxis. Both Thompson and Cook comment at length on the spiral inflorescence of the petals, florets, buds, fruits, leaves, spines, or scales of such plants as the cauliflower, the daisy, the giant sunflower, various succulents, and the cones of many firs. Often, an intricate and precise pattern of lozenges is created by the intersection of two oppositely directed logarithmic curves, called "*dia-dromous* spirals" (TG 916), as exemplified by "the curving rows of florets in the discoidal inflorescence of a sunflower" or "the crowded assemblage of woody scales" on "the surface of a pine-cone" (TG 913).

Among the fauna, there is the spiral growth pattern of horns and tusks, nails and claws, beaks and teeth,

> where about an axis there is some asymmetry leading to unequal rates of longitudinal growth, and where the structure is of such a kind that each new increment is added on as a permanent and unchanging part of the entire conformation. . . . The logarithmic spiral *always* tends to manifest itself in such structures as these, though it usually only attracts our attention in elongated structures, where (that is to say) the radius vector has described a considerable angle. (TG 896–97)

While "the incisor tooth of a rabbit or a rat" may seem straight, this is only because the short length of the specimen obscures its logarithmic growth curve.

Finally, there is a full range of behaviors that involve a "spiral turn," such as the spiralic path traced by a spider spinning a web, the coiling of a serpent, or a dog's "spiral movement before lying down to sleep" (MM 49). In the case of the predatory spider and snake, the spiral serves as a strategy for survival, enabling the one to construct efficiently a trap for its prey and the other to stun a victim or defend itself by concentrating and storing a large amount of energy—the spiral coil conduces readily to this—which may then be released in a sudden, lethal, self-propelling strike.[6] In the case of the dog, the turning round not only serves to hew out a comfortable niche in, for example, a bed of leaves but also curls the animal's body in a manner that will protect vital organs and conserve body heat.

The Helix

There is a three-dimensional extension of a spiral-related form, frequently referred to as a spiral, which I intend to distinguish by the term "helix." Actually, the word ἕλιξ in Greek, from the verb "to turn round," is roughly synonymous with the Latin *spiralis,* from *spira* ("coil"), and the words have been used interchangeably and often arbitrarily to designate any number of twisting or spinning phenomena. The distinct definitions I propose assigning them are not, however, without precedence in past and current usage. Specifically, the spiral remains the two-dimensional, discoid, coiled figure previously described, while the helix exists only as a three-dimensional form[7] and represents a record of a progressive circular spin about an axis. In its "geometrical" definition, the OED refers to "helix" as three-dimensional and thus "distinguished from *spiral* which is applied only to plane curves" (1285).

The reason that the helix, as envisioned here, is exclusively three-dimensional is that its corresponding plane figure is a simple circle, from which it can be generated by extending the circle's center in a straight line, perpendicular to the circle's plane. When confined to the original plane, revolution of the end point of a radius along the circle's circumference is unvaryingly repetitive and an example of "static" motion. If the radial line is not only rotated but also translated along the central axis line at a constant rate, the figure generated by the end point of the radial vector will be helical. Seen from another perspective, the helix's successive generation of coils, equidistant from the axis, if compacted tightly, would produce a cylinder, and for this reason, what I term simply a helix is sometimes called a "cylindrical helix," in order to distinguish it from the conical variety, which will be examined next.

First, the peculiar properties of helical rotation must be discerned. Unlike the repeated revolutions of two-dimensional circular motion, which are, in Heideggerian terms,[8] "the same" or "tautological," the successive revolutions of the helix may be "identical," i.e., "equal to" one another, but not "the same," because the return to the beginning of the circuit will be at a point parallel to, but not the same as, the original point of departure. Helical movement is thus "cyclical," not simply circular, and involves "periods" of revolutions along a continuum, like the breaking of a wave at the seashore in a horizontal, cylindrical-helical progression.

Because the amplitude and frequency of such a helix, and of most helices, tend to be uniformly identical or similar, if nonetheless progressive and persistent, helical motion would seem to imply a monotonous recurrence of duplicate cycles over time:

> Unlike the spiral, the helix doesn't swing steadily wider from a point of origin. Monotonously, the helix just strings along always at the same diameter. That turns out to be the secret of its

> importance. The helix is oddly rare in inorganic nature. In living things, it appears frequently and fleetingly in the forms we see—perhaps in a crisp curl of hair or in the grasping tendril of a vine. The domain of the helix is among the processes of life at the smallest level where we can speak of such processes at all—at the level of the individual molecules that determine all that goes on within the living cell. (JS 37–38)

The allusion here is to the "helical model of amino acids—the alpha helix—that proved to be frequently found in the structure of proteins," as well as to "the structure of the genetic material itself" (JS 40), namely, the double helix of deoxyribonucleic acid.

Actually, there are many common, everyday objects and phenomena that display a helical structure. Some of these are misnamed, such as the "spiral binding," which is actually helical, and the "spiral staircase," which is more accurately signified in French by the term, *escalier en hélice*. On the other side, there are so-called helices that are principally spiralic, such as the common snail (*helix hortensis*)[9] and the volute, which in architecture is sometimes referred to as a helix, even though, as has been shown, it is actually a spiral.

Two helical objects, the corkscrew and the drill bit, demonstrate how the helix is useful mechanically, not only for boring into certain materials but for moving the extricated matter out along successive helical convolutions in the manner of a conveyor, just as Archimedes' famous screw was able to raise water. Another common object, the helical spring, exhibits, like the spiral spring, the mechanical property of certain materials to store energy through contraction, which is then available for immediate release. The capacity of a steel spring, for example, to contract under pressure and to return to its original state when that pressure is removed makes it useful for absorbing shocks (e.g., the elasticity of furniture springs), for "triggering" mechanisms (e.g., the "automatic" release of a cocked gun hammer), and for determining weight (e.g., the displacement measured by a spring scales).

Other notable examples of the helical design are the human umbilical cord, "composed of two arteries and one vein twisted into a triple helix, connecting mother to child" (SL 142), any number of similarly braided strands, climbing plants that wind helically about a host stalk or trunk, and propellers, including those that move horizontally, like pre-jet aircraft, and those with a vertical thrust, like the appropriately named "helicopter." Finally,

> the planet earth itself displays a helical movement within the overall galactic unit. When we consider the earth's movement, we usually think of its rotation on an axis and its revolution around the sun. But the sun, too, is moving along with it. When we add this characteristic of the earth's movement to the other two, we discover that the planet designs a helical path through space. (SL 30–31)

The Spiral Helix

Although the spiral and the helix have now been clearly distinguished, there still remains one further expression of spiro-helical orientation, which exhibits both the alternately increasing or decreasing circular movement between center and circumference of the spiral and the helix's translation along a third-dimensional axis. Viewed first from the helical standpoint, the successive circlings, instead of being generated uniformly and infinitely along a continuum, now spiral out from or in toward a single point of origin or completion on the axis about which they turn, giving the figure the asymmetrical properties of both closed and open ends, and if the coils were compacted, they would yield, not a cylinder as does a helix, but a cone. Alternately, from the point of view of the spiral, the convolutions do move progressively from center to periphery (or vice versa), but rather than aligning themselves in a single plane, they telescope in or out (depending on direction), revolving about a third-dimensional axis, as when the locus of a point rotates at a constant angle round a cone. A hybrid, then, of both helical and spiralic characteristics, it is fittingly termed a "spiral helix."[10]

Specifically, the difference between a bolt and a screw, a corkscrew and a gimlet, illustrates the difference between a helix and a spiral helix. The bolt and the corkscrew are helical because they unfold cylindrically; the screw and the gimlet spiro-helical because they culminate conically in a single end point or tip. What is referred to as the "vegetal helix" (SL 32–33) is thus technically a spiral helix, since, as Thompson points out, the offshoots, although aligned helically around the trunk or stem, wind centripetally to a continuously "evolving" tip-probe of new growth. But more obvious examples of spiro-helical structuring can be found in marine phenomena, such as the turbinate seashell, the important symbolic properties of which are analyzed at length in Chapter 1.

The Gyre

At this point of transition between the essentially static, abstract geometrical structures just analyzed (where motion is only implied) and the dynamic whirling phenomena I intend presently to take up, a brief word concerning the "gyre" seems appropriate, since it is a term applied rather indiscriminately to various forms of turbulence. From the Greek γῦρος, "ring" or "circle," its definition also includes "spiral" and "vortex" (where the emphasis is upon structure), as well as "a turning round, revolution, whirl; a circular or spiral turn" (where types of motion are stressed).[11] For Dante, the circular connotations of *giro* and *giri* seem paramount (see Chapter 5). In the poetry of W. B. Yeats—he pronounced the word with a hard "g"—it refers especially to the spiro-helical pattern and is linked to the unusual word "perne," as in the phrase "perne in a gyre" of "Sailing to Byzantium," where the verb, fashioned out of the noun "pirn" (bobbin) by the poet himself, denotes a helical "winding" or "unwinding." In all, the word "gyre" seems principally to be a poetic and literary term, as the OED suggests, and its use is rare. By contrast, "gyrate" and "gyration" are common words, derived from the same Greek root, that betray the now-predominate dynamic connotations of these and of the following signifiers of turbulence.

The Vortex (Tourbillon)

When one turns to "the literature of the vortex" in search of a definition, it is soon apparent that this phenomenon is of particular interest to physics, not only because it occurs frequently in nature (φύσις) but because turbulence, defined by one source as "randomly distributed vorticity,"[12] "remains one of the last great problems of classical physics"

(JS 15). Generally, a vortex is defined as "a mass of fluid in which the flow is circulatory" (VN 1784) or "fluid motion involving spin about an axis"[13] or, in the case of the French synonym *tourbillon*, "type of flow characterized by a rotating movement of fluid particles about an axis, with a speed inversely proportional to the distance from the axis."[14] The qualification of varying velocities, in this last example, is actually characteristic of a particular type of vortex, and it is thus necessary to take a brief look at vortical morphology.

Rotational fluid motion "in which the velocity varies inversely as the radius, is referred to as a free vortex. . . . The tornado is an example of a free vortex, with high velocities near its center, and correspondingly low pressure intensities. The waterspout is its counterpart over water."[15] In a free vortex, "the stream lines are concentric circles"[16] with velocity increasing centripetally (toward the center) and decreasing centrifugally (toward the periphery). There are actually two categories of "free" vortices: the "free circular vortex" just described, consisting of concentric circular streamlines, and the "free spiral vortex," in which "the stream lines will be logarithmic spirals. When water is delivered from the circumference of a centrifugal pump or turbine into a chamber, it forms a free vortex of this kind. The water flows spirally outwards, its velocity diminishing and its pressure increasing" (EB 45).

Opposed to these free forms, which occur freely in nature, is the "Forced Vortex . . . in which all the particles have equiangular velocity" (EB 45) and "the fluid rotates as a solid body" (TE 669). The uniformity of velocity across the streamlines occurs, in this case, as the result of an additional force, such as "radiating paddles revolving uniformly": "If the law of motion in a rotating current is different from that in a free vortex, some force must be applied to cause the variation of velocity" (EB 45).

Research into the incidence of naturally occurring vortices leads one either to the study of hydraulics (hydrodynamics), for which maelstroms, whirlpools, and eddies are the prime examples of scrutiny, or to the study of aerodynamics and meteorology, which are concerned with "airfoils," "intrados and extrados," and "trailing vortex sheets"—terms too technical for the purposes of this study—as well as whirlwinds, hurricanes, cyclones, typhoons, tornados, waterspouts, sandspouts, and dust devils. The latter group of primarily air-related phenomena is usually characterized by "a double movement of translation and rapid rotation,"[17] which accounts for the wide ranging, destructive mobility that they are both capable of and infamous for. However, there does exist in both air and water a "stationary vortex" (VN 1784) or "bound vortex,"[18] i.e., "A vortical motion steady in space and time. The standard example is the two stationary eddies found behind a cylinder placed in a stream of viscous fluid. Another is the bound vortex trailing from the wing-tip of an aircraft" (VN 1784). One need only observe the downstream side of a bridge pylon to find a common manifestation of these stationary swirls.

Although Helmholtz, Poincaré, Karman, and others have identified certain properties of vorticity, neither the causes nor the mechanics of vortical motion are fully understood. Regarding the formation of cyclonic storms, one source concludes,

> It is probable that the cause of *vortices* is highly complex; that one must look for it in a concurrence of circumstances deriving at once from the meteorological constitution of tropical zones and from the pattern of winds during the warm season in these regions. But these are vague conditions from which it is impossible to adduce a precise explanation. (LG 357)

Other investigators echo this uncertainty concerning the causes:

> Two principal theories have been presented to explain the formation of great vortices: one thermal, the other mechanical, the latter seeking the cause in the meeting or superimposition of atmospheric currents, which also develop from solar activity. In spite of the great talent deployed by their authors, the one and the other still seem, in part, premature.[19]

As baffling as the unknown etiology of certain vortices are conflicting observations and interpretations of what exactly occurs at the center of such a swirling mass. The "eye of a hurricane" is a rather overused metaphor for tranquility in the midst of turbulence, but in nature, the center of such storms, although often still, is yet the region of highest wind velocities:

> At the center of the meteor, there usually reigns a relative calm, sometimes even an absolute calm, attributable to the rarefaction of the column of air around which the cyclone turns like an immense ring: this is the axis or eye of the storm. It is not rare at this point to see the clouds dissipate and the azure or the stars of the sky appear

> an instant. It is however near this center that the force of the *vortex* is to be feared the most. (LG 356)

In the case of a tornado or waterspout, each "an example of a free vortex, with high velocities near its center, and correspondingly low pressure intensities" (MH 431), no stillness is apparent, but these are vortices on a much smaller scale than hurricanes, and evidence of unscathed, untouched areas in the midst of total destruction are frequently reported after tornados have ravaged a region, suggesting that the smaller whirlwinds may have the same paradoxical properties as the great ones. Finally, sightings of even stranger occurrences in the middle of cyclones have been reported, such as the following account based on observations by the English engineer, Piddington:

> On passing through the center, a formidable noise resembling artillery rounds, a continual rumbling of thunder explodes and dominates everything. Near this center, where the greatest emptiness occurs, the wind, rising up, seems to describe an immense spiral. Its fury redoubles. At the axis of the cyclone, a powerful suction raises the sea into a conical mountain and forms the storm surge which, advancing over the Ocean's surface, floods the coasts and produces the terrible phenomenon of tidal waves. (LG 357)

It should not be surprising that such violent events are apt to occur near the center of a vortex, since, as has been shown, low-pressure intensities (which account for rapid displacement of particles through funnel suction), high velocities, and rarefaction are all states that pertain to the vortical core. In theory, the velocity should be infinite, but the effects of inertia, friction, drag, and other properties of terrestrial viscosity inhibit any "pure" realization of the principle and produce certain anomalies, such as the vortex filament, a central tube of fluid which is, in the case of the free vortex, a "rotational" axis about which the "irrotational" particles of the surrounding fluid whirl:

> It is clear that the infinite velocity at the centre of the free vortex could not occur physically, and in real flows the centre of a free vortex must be occupied by a solid cylinder (with an associated boundary layer) or by a core of fluid rotating in a physically possible manner. A convenient idealization is to consider the core

to be a forced vortex. Such a core contains all the vorticity of the vortex and is called a vortex filament. (TE 669)

One of the "laws governing vortex flow" attributed to Helmholtz states that "filaments have no ending. They are either closed paths, or the ends extend to infinity" (VN 1784). While the "closed paths" of the circular vortices called "vortex rings" are realizable in nature, occurring as "smoke rings or ground explosions," for example, other "infinite" manifestations are clearly not possible in our finite earthly milieu. In the extreme conditions of the void, however, perhaps in distant galaxies (or even at the center of our own), where forces yet unknown prevail, all implications of vortex theory may be capable of the fullest actualization, as some recent scientific speculation seems to imply, most particularly in regard to the "entirely theoretical objects"[20] now referred to as black holes.

A black hole is a cosmic vortex that results from the collapse of a star greater than three solar masses. Einstein's theory holds that, in such a case, the force of gravity will overcome the resistance of degeneracy pressure, resulting in "a singularity in the hole's center, a point at which matter is crushed to infinite density and zero volume" (SB 64). The gravitional pull of such a locus would be so great that even light would be sucked toward the center, producing the "ultimate vortex" from which nothing passing within a certain range could escape.

Although only indirect evidence to date supports empirically the existence of black holes, the likely properties of such an entity have been deduced from implications of theoretical physics and can be demonstrated, as Shipman has done in the thought-experiment he conducts based on the contraction of a "ten-solar-mass object" into a hole "60 kilometers across" (71). I think it worthwhile envisioning, for a moment, a journey into a black hole, precisely because it remains as yet a transcendental construct, akin to the symbols I am analyzing in this study. Such a mental leap conveys us, in effect, to the limits of Einstein's universe, where ultimate principles of vorticity take over, ushering beyond into inverted anti-universes of pure imagination.

The journey begins with a probe venturing into the region of a black hole, where it would experience "a weak but relentless gravitational pull" (67) as it began to orbit about the invisible object, causing it to spiral gradually toward the vortical center. The nearer it approached, the greater the effect of "tidal forces" would be on the probe itself, squeezing and elongating it. Simultaneously, time would begin progressively to slow down, such that the ticking of a clock on board the vehicle would "take place in slow motion" (72). Upon reaching the "event horizon" or "point of no return," from which nothing could escape due to forces now approaching infinity, "it would take an infinite time until the next clock tick. Events would be frozen. Time comes to a stop at the event horizon" (74). Here, the perspective is that of an outside observer stationed nearby the hole. The experience of time "relative" to the probe, now inside, would be quite different. It would pass through and "reach the center 67 millionths of a second after it passed the horizon" (79). In so doing, it would disintegrate and merge into the "singularity" at the hole's center, a paradoxical point where, as has been suggested, matter is so dense that it has no measurable dimensions: "A singularity is an absurdity. It is a point containing all the mass of the hole. The singularity has zero volume, and the density of matter is infinite. The tidal forces are infinite. So the theory says, anyway" (79).

The threshold of a singularity is "where Einstein's theory of gravity breaks down as the forces of gravity take off towards infinity" (82). Whether or not, at this point, "wormholes" burrow into hypothetical other universes, which, conversely, spill back into our own through "white holes," is a matter of conjecture on the fringe of astrophysics; such "space-warps" are, for now, beyond the grasp of scientific description and certainty. But they do offer an almost irresistible appeal to the imagination and are pregnant potential symbols of ultimate destruction and transcendence, comparable to the most powerful traditional symbols of cosmic turbulence and metamorphosis.

Complementarity, Enantiomorphism, and Handedness

The four terms announced in the title of this appendix have now been accounted for, although I have limited my analysis so far to unique (singular) representations of each configuration. Because we are dealing in most cases with asymmetrical forms, a complementary dualism is inherently implied (and often manifested), necessitating a brief discussion of "enantiomorphic" pairs.

Enantiomorphism is characteristic of asymmetrical forms that possess what we commonly call "handedness," or, to quote Martin Gardner's fascinating study, *The Ambidextrous Universe,* "Two asymmetric figures, each the mirror image of the other, are said to be *enantiomorphs*. Each is enantiomorphic to the other."[21] Most people take left- and right-handedness for granted and never stop to consider the many subtleties, complexities, and paradoxes of these seemingly obvious indications of position or direction, especially in cases of the types of rotational motion that are the focus here.

Right and left are clearly anthropocentric terms derived from the body's bilateral symmetry along the plane of symmetry that bisects the head and trunk into left and right sides. But the human body also has an asymmetrical "front" and "back," which means that "right and left involve, not merely bilaterality, but *dorsi-ventrality, fore-and-aftness*" (CC 263). The relative nature of right and left is quite clear when I face another person or my own reflection in a mirror. My left is the other person's right; if I wave my left hand at my mirror image, my right hand waves back at me.

An asymmetrical figure such as a spiral, similarly reflected in a mirror, reverses both its handedness and the direction of spiralic gyration, so the two figures are enantiomorphs. Enantiomorphism is, in fact, an inherent property of all asymmetrical structures, from molecules that display "racemic form" (i.e., "an equal mixture of left- and right-handed molecules" [GA 122]),[22] to the reverse handedness of Siamese twins, which "are exact enantiomorphs in almost every detail" (GA 74), or the mirror imagery of "conjugate" (oppositely turning) pairs of animal horns. It is nonetheless curious that many potentially racemic structures manifest a distinct tendency to exist in one or the other handed form but not in both. Most discoid and turbinate seashells, for example, turn in one direction and only rarely "the other way." Various species of twining plants twist variously to the right or left, but a given species tends to favor one direction for reasons not fully known, although the problem of torsion in climbing plants has been studied by many great scientists, including Charles Darwin. Then there is the extraordinary fact that "all amino acids in living organisms" display one-handedness, while "all helices of protein and nucleic acid" are oppositely handed (GA 183).

Now, the question arises, which of these turn to the right and which to the left? Here, as Thompson, Gardner, Cook, and others regretfully admit, the literature is hopelessly confused. A simple thought-experiment will demonstrate why. Suppose you wish to determine whether a certain helical stair winds to the right or left. If, while ascending, the central axis (newel) is always to your left, you might reasonably conclude that it is "leiotropic." But descending, the newel will be on the right, so the same stair is then "dexiotropic."[23] The point is that, in the case of the helix, direction of rotation is reversible, depending on which end is the starting point. The stair is either rightward or leftward, depending on the perspective of the viewer, but is itself inherently neither (or both). In the case of a spiral or a spiral helix, the direction of gyration is also reversible, depending on whether the movement is from periphery to center or vice versa.

It is not enough to say that a given spiral, helical, or spiro-helical figure is simply left- or right-handed, even though this has traditionally been done, resulting in conflicting interdisciplinary taxonomies. Cook's attempt to set a standard, based on the conchologist's practice of classifying the majority of shells as right-handed (even though the direction of centripetal circumvolution is "leiotropic" [26]) is unfortunate and unacceptable, since it leads to the equation of clockwise motion with left-handedness and counterclockwise motion with right-handedness (39), the very opposite of what we commonly understand. A clock's direction, like that of the sun (viewed in the northern hemisphere), has traditionally been considered "to the right," and in this study, right-handed (dextral, dextrorse, dexiotropic, dextrorotary) is equated with clockwise movement and left-handed (sinistral, sinistrorse, leiotropic, levorotary) with counterclockwise circumvolution.[24]

It is important to maintain these ancient directional associations, because, on the symbolic level, they carry with them deeply engrained positive and negative connotations. Mackenzie devotes an intriguing chapter to these contrary movements entitled "The Sacred Circuit" (123–38). From ancient times, sunwise rotation, referred to as "making the dessil" (from the Gaelic, *Deiseal*), has been considered "the proper direction," whereas left-handed or "backward" movement counter to the apparent direction of the sun, called variously "widdershins," "withershins," "witherwise" (also, in Gaelic, *Tuaitheal* or *tuathuil*), has been regarded as "ominous" and "unlucky" (123–25).[25]

> The ceremony of encircling sacred objects and places was sometimes a spiral movement. It is of widespread character and great antiquity and appears to have originated in the belief . . . that it was necessary to stimulate the Great Bear to revolve in the right direction at the beginning of the new year, and also at the beginning of each season. By following the course of nature, the magic-

> workers not only assisted nature, as they believed, but procured for themselves an accumulation of "good luck"—that is, everything man desired, health, food, prosperity, etc.; by inverting the course of nature the magic-workers either performed a ceremony of riddance or invoked the forces of evil. (124)

Among examples of such ritual circlings mentioned by Mackenzie are Vercingetorix's act of submission to Caesar, which entailed "circuits" on horseback performed by the former about the Roman conqueror, and St. Patrick's consecration of Armagh, which culminated in a walk sunwise about the site (126). Cook even claims that "Port decanters . . . must follow the sun [and] . . . we still waltz 'with the sun' " (165–66). Nevertheless, the attribution of good or bad luck to direction of circulation is a product of symbolic association that occurs on the hermeneutical level, since no such quality can be said, logically, to inhere in vectorial direction itself.

With these important distinctions concerning handedness understood, we are ready to look at specific cases. Among examples of enantiomorphic spirals in nature and art, conjugate pairs of animal horns have already been mentioned.[26] The spiral volute of classical Greek architecture also occurs in oppositely rolled pairs on the capital of the Ionic column. As for enantiomorphic helices, the "dextrorse" (twining upward to the right) and "sinistrorse" (twining upward to the left) growth of vines has been alluded to, of which the honeysuckle and the bindweed are, respectively, ascending clockwise and counterclockwise examples. A further striking example would be the beautifully wrought pair of oppositely twisting bronze and white marble elliptical helical staircases, ascending five stories on each side of the Courtroom of the United States Supreme Court Building.

Currently, one of the most fascinating instances of helical complementarity is the sculpturesque "double helix" of DNA, which Watson and Crick first discovered in March 1953 and described as "right-handed with the two chains running in opposite directions."[27] More specifically, the molecule consists of

> two repetitive strands, one winding up, the other down, but hooked together, across the tube of space between them, by a sequence of pairs of chemical entities—just four sorts of these entities, making just two kinds of pairs, with exactly ten pairs to a full turn of a helix. (JS 10)

The complementarity of this structure resides in the reversed sequence of the bases adenine, cystosine, guanine, and thymine, as well as in the linking of the A—T and G—C pairs themselves, not in the direction of helical rotation, which is the same for both strands. Thus the true enantiomorph of the normal DNA molecule would be an oppositely handed double helix, such as researchers at M.I.T. claim to have discovered, which indicates that "the double helix can be flipped over . . . by the action of certain chemical sequences along the rungs—or by chemical conditions outside them"—from the normal smooth pattern to a form "twisting jaggedly" in the opposite direction.[28]

Another figure that involves a double helix, although in this case it is really a question of a conical spiral helix, is the "hermeneutical" caduceus or magic wand of the messenger-god Hermes, which consists of a winged staff with two serpents twined about it.[29] At first glance, the two snakes seem to be antitheses, since they do wind about the axis always at opposite positions (180 degrees) of the expanding conical circumference. Close scrutiny reveals, however, that, like the dual strands of DNA, both serpents twist in the same direction, which in all but one of the illustrations I examined ascended to the right. Although faced off one against the other, they are not in fact enantiomorphic pairs.[30]

It is clear, then, that the notion of oppositeness is more complex than it might first seem, and this is particularly true in the case of asymmetrical figures like the spiral. A symmetrical object such as the circle does not, in and of itself, suggest a complementary "other," and it is no wonder that the circle is often a symbol of unity, *mono*tony, and changelessness. A simple degree of antithesis may be added to the form by means of hue (e.g., dark versus light) or direction of circumvolution (clockwise versus counterclockwise), or the circle may be divided into complementary halves. In the case of asymmetrical figures like the logarithmic spiral, though, the form seems already to be a half that necessitates an other half for it to be complete or "unified." And since it contains latently within the very singleness of its structure that reversed other half that is its double, it is an example par excellence of dialectic, in which a principle of reversible oppositeness inheres. There is oppositeness of orientation, like the two magatamas of the Tao symbol ("right side up" versus "upside down"); there is oppositeness of the circuit vector (rightward or leftward rotation); and because the spiral embodies a center and a periphery, there is an oppositeness of "in" and "out," that is, of centripetal movement toward the center (a yang tendency in Oriental symbolism) or of centrifugal movement away from the center (a tendency of the yin). In the case of the three-dimensional spiral helix, this also involves translation along an axis.

Phenomena that exhibit the fullest range of such dialectical complementarity and complexity tend to be found in "fluid flows" and particularly within the domains of aerodynamics and hydrodynamics. Of all the spiral and spiro-helical elements we have considered so far, the dynamic vortical manifestations of the principle have been associated primarily with fluid states. Air and water are "elemental" fluids, and of the other two traditional elements, fire, which subsists within the medium of air, also displays properties of a fluid and is related directly and tangentially to vortical motion (for example, the vortical smoke ring, which is "tangential" to the flame itself, is a closed circular vortex of "concentrated vorticity" [VN 1784] in which the filament "is closed on itself to form a ring" [TE 670]). As for the fourth element, earth, such solid vortices as wood whorls or seashells are actually "solidified" substances, since they are vortices from which all viscosity seems to have receded, although their original dynamic formation is indicated by the vortical configuration itself, which represents "the record of a spin." (The "fluidity" of solid as opposed to liquid forms is, in this sense, relative, like the magnitude of difference between geological and diurnal time or cosmic and terrestrial distance.) All of the elements, then, exhibit varying degrees of vorticity, but in a manner that seems to conform to the following general rule: the greater the fluidity, the higher the incidence and the greater the complexity of vorticity.

In aerodynamics, the problems posed by turbulence about an airfoil testify to a very high degree of vortical complexity, and some, but perhaps not all, of the properties of dialectical and/or enantiomorphic spiral helices are discernible. The distinctive shape of an airfoil, i.e., the asymmetry of its intrados and extrados, gives rise to a complex interplay of "bound" and "trailing" vortical effects, which produce the two oppositely rotating "tip" vortices that trail downstream from an aircraft and constitute what is referred to as the downwash. Complementary vortical interaction is thus integral to the concepts of lift and drag that permit flight.

A hydrodynamic example of a similar type of turbulence, also caused by the interruption of fluid flow by a solid barrier, is the "vortex sheet" that forms in the wake of pylons supporting a bridge. The "path of vortices" in this case seems to conform to the system of alternate whirls ("vortices arranged in quincunx" [GL 10:409]) described by Karman, although there is another type of vortex sheet, formed by the slippage of one layer of fluid over another, in which the vorticity is uniform in the direction of surface flow (see MH 431).

But the vortical phenomenon with which I shall conclude this examination of complementary spiral helices—as well as, for that matter, the full survey in this Appendix of spiralic, helical, and vortical phenomena—is the peculiar effect involving air and water near the equator of a revolving sphere (specifically the earth) described by the famous theorem of Gaspard Gustave de Coriolis, a mathematician at the Ecole Polytechnique of Paris in the first half of the nineteenth century known also for his analysis of the mechanics of billiards. Accordingly, "the rotation of the Earth deflects all the air currents towards the right in the Southern hemisphere" (GL 3:508). Isaac Asimov[31] adds that deflection toward the east is true of both air and water currents moving away from the equator (elsewhere termed "descending" [GE 234]), just as deflection toward the west is the case for currents moving toward the equator (termed "ascending"). From this perspective, the equator is not only the dividing line for opposite vortical effects,[32] but also a zone of complex, contrary turbulence and interplay.

Certainly, if spiro-helical enantiomorphs are to be found, it is in just such "zones of interaction" (GL 10:409), as the following complicated classification of atmospheric vortices by Durand-Gréville seems strongly to suggest:

> There are three kinds of atmospheric vortices: 1. *Fixed permanent vortices:* a. descending, centrifugal, turning from left to right in the N. hemisphere, from right to left, in the S. hemisphere . . . ; b. ascending, centripetal, turning from right to left in the N. hemisphere, from left to right in the S. hemisphere. . . . 2. *Fixed temporary vortices:* a. descending, centrifugal, covering all the continents of the N. hemisphere during our winter, those of the S. hemisphere during their winter . . . ; b. ascending and centripetal, covering our continents during our summer, those of the S. hemisphere during their summer. . . . 3. *Traveling vortices:* a. descending and centrifugal, the whole length of the two tropics, merging with the W. side of the permanent anticyclone that they meet, and emerging by scission from the E. side; b. ascending and centripetal, traveling toward the W. between the tropics (cyclones, hurricanes, typhoons), toward the E. in the temperate zones (squalls, barometric depressions); c. ascending and centripetal, like the latter, but of

> very small diameter,—from 1 m. to a few hundred meters,—traveling with the depression in which they are but a minute perturbation (vortices of heat, waterspouts, tornadoes). (GE 234)

My purpose in recording these categories is both to demonstrate the intricacy of turbulence itself and to elicit the diverse qualities and degrees of vortical complementarity and inversion that the Coriolis force engenders. Suffice it to say that the preceding classification describes various combinations involving polarities of clockwise and counterclockwise, centrifugal and centripetal rotation; of translation across the Earth's surface in northerly, southerly, easterly, and westerly directions; and of seasonal (thermal) as well as barometric intensity.

This inquiry into the vocabulary of turbulence culminates fittingly with an allusion to the Coriolis effect, which illustrates by its very complexity the problems of terminology and taxonomy that any study of turbulence must necessarily confront. I have attempted, in this Appendix, to evoke the sweeping range and magnificence of turbulent phenomena but also (and more importantly) to identify and define the spiral, helical, spiro-helical, and vortical paradigms that underlie them, making it possible to grasp and distinguish them, so that the symbolic structural analysis of turbulence, which is the main thrust of this book, is established firmly upon a clearly articulated system of signs. I shall be content, in any case, if I have succeeded in bringing a measure of logic and uniformity to a field that, when I began my research, seemed hopelessly confused, inconsistent, and contradictory.

Notes

Chapter 1

1. Nahum Stiskin, *The Looking Glass God . . . Shinto, Yin-Yang, and a Cosomology for Today* (Tokyo: Autumn Press, 1972), 116–17.

2. Jean Chevalier and Alain Gheerbrant, *Dictionnaire des symboles* (Paris: Laffont, 1969), observe that "*The circle adjoined to the square is spontaneously interpreted by the human psyche as the dynamic image of a dialectic between the transcendent celestial to which man aspires naturally, and the terrestrial where he is actually situated*" (159).

3. See Jill Purce, *The Mystic Spiral, Journey of the Soul* (New York: Thames and Hudson, 1980), for extensive illustrations of spiral symbolism.

4. Donald A. Mackenzie, *The Migration of Symbols* (New York: Knopf, 1926), 48–49.

5. Mackenzie does not distinguish between the spiralic (discoid) and spiro-helical (turbinate) forms that have been differentiated by conchologists. Seashells display a range of variations between both extremes. The nautilus, for example, tends to uncoil in a plane more or less spirally, while a conch, such as the triton, translates spiro-helically along an axis.

6. Mircea Eliade, *Images et symboles* (Paris: Gallimard, 1952), 171.

7. Gaston Bachelard, *La Poétique de l'espace* (Paris: Presses Universitaires de France, 1978), 106, citing Paul Valéry.

8. Mackenzie alludes to Hesiod's account of the castration of Saturn by his son Kronos and the birth of Aphrodite amidst "whirling water" (71) from the severed organs. Botticelli's depiction of the allegory in his *Birth of Venus* indicates that by the time of the Renaissance the "shell birth" had become the dominant variant.

9. Bachelard citing Edouard Monod-Herzen. When transposed through translation from French into English, the term *hélices spirales* corresponds to my own definition of "spiral helices" (see Appendix).

10. The shell that contains the living creature is itself dead, eliciting the following observation of D'Arcy Thompson, *On Growth and Form*, II, 2d ed. 1942 (rpt. London: Cambridge University Press, 1979): "The logarithmic spiral is characteristic, not of the living tissues, but of the dead" (767).

11. The vessel's shape, moreover, is mirrored structurally by the ear to which it is held. Actually, the "ear helix" is a spiral helix like the seashell, and this murmuring chamber may well have corresponded to the mythological "ear of chaos." According to Mackenzie, ancient texts and artifacts indicate a wide-

spread belief that the "breath of life" (143) entered through the ear, making it, like the female vulva (and the *coquille* itself), a symbolic orifice of procreation. In this respect, the seemingly ludicrous birth of Rabelais's Gargantua via his mother's left ear canal, while obviously intended for comic effect, is nonetheless consistent with a very ancient tradition of ear symbolism.

12. The verb is related to the nouns *turba* ("tumult . . . disturbance, commotion") and *turbo* ("a whirling round"). *Cassell's Latin Dictionary,* by D. P. Simpson (New York: Macmillan, 1968), 618–19. *Turbo* can also designate "a whirlwind or tornado, a spinning-top, a reel or spindle, a whirl, twirl, twist, revolution." *Oxford English Dictionary [The Compact Edition of the]* (Oxford: Oxford University Press, 1979), 3434.

13. L. T. Hobhouse, *Morals in Evolution* (London: Chapman and Hall, 1951), 187n., citing "Griffith's trans. following Erman, *World's Literature,* p. 5340."

14. Isaiah 5:28, 66:15; Jeremiah 4:13.

15. Psalm 58:9; Proverbs 1:27, 10:25; Jeremiah 23:19, 30:23; Amos 1:14; Zechariah 7:14.

16. Isaiah 41:16, on page 668 of *The Dartmouth Bible* (King James Version), ed. R. B. Chamberlin and H. Feldman (Boston: Houghton Mifflin, 1961); other examples include Nahum 1:3 and Jeremiah 9:14. Specific biblical quotations are from this edition of the Bible.

17. Isaiah 17:13, 40:24; Hosea 8:7.

18. Job 38:1 and 40:6 (B 457 and 459).

19. The view expressed in *A Dictionary of the Bible,* ed. James Hastings (New York: Scribner's, 1905), that the whirlwind describes a "physical" state of "rapture" in the cases of Job, Elijah, and Ezekiel (IV, 915) seems quite plausible as far as it goes but fails to discern the particular nature of the symbolism and its depth in each instance.

20. For Blake, as we shall see, they symbolize the "Four Zoas," i.e., the four basic factors of every personality: imagination ("Los"), emotion ("Luvah"), intellect ("Urizen"), and sensation ("Tharmas").

21. Plato, *Timaeus,* trans. B. Jowett, ed. G. R. Morrow (Indianapolis: Bobbs-Merrill, 1949), 20. Plato also affirms the rectilinearity of movement on the earth to which he assigns "the cubical form" (38) and the six inferior or "deviating motions" in three-dimensional space, viz., "up, down, to the right, to the left, forward, backward" (16).

22. Homer, *The Iliad of Homer,* trans. Richmond Lattimore (Chicago: University of Chicago Press, 1974), XII:40 (259). For the sake of consistency, I intend to observe the English spellings of Greek names as they appear in the translations used, with the exception of "Achilles," which the weight of tradition renders preferable to Lattimore's—albeit more authentic—"Achilleus."

23. XXIII:13–14 (450) and XXIV:14–18 (475).

24. Homer, *The Odyssey,* trans. Robert Fitzgerald (Garden City, N.Y.: Doubleday, 1963), 61.

25. See Mackenzie (127) for examples of such ceremonies in Greece. As the direction of the circuits is not specified in this case, the purpose of the ritual cannot be determined.

Chapter 2

1. Michel Serres, *La Naissance de la physique dans le texte de Lucrèce: fleuves et turbulences* (Paris: Editions de Minuit, 1977), 9.

2. Another manner in which the Pre-Socratics anticipate Plato (and in particular, Platonic idealism).

3. Unless otherwise noted, the dates of the Pre-Socratics are those of G. S. Kirk and J. E. Raven, *The Presocratic Philosophers: A Critical History with a Selection of Texts* (Cambridge: Cambridge University Press, 1977).

4. Aristotle, *Metaphysics,* trans. W. D. Ross, *The Basic Works of Aristotle,* ed. Richard McKeon (New York: Random House, 1941), 693–94 [983b].

5. Kathleen Freeman, trans. and ed., *Ancilla to the Pre-Socratic Philosophers: A Complete Translation of the Fragments in Diels, [H.], Fragmente de Vorsokratiker* (ed. with additions by W. Kranz) (Oxford: Blackwell, 1948), 27.

6. KR 199n. I see no reason to duplicate the authors' practice of italicizing their translations.

7. Later "prester" in Lucretius's *De Rerum Natura,* as we shall see.

8. FD 28, fr. 49a. KR are doubtful of the authenticity of this particular formulation (see 198 n. 2), but they do accept Diels fragment 12: "Upon those that step into the same rivers different and different waters flow" (their translation, 196).

9. The idea of "four elements" is a product of "post-Empedoclean speculation" (KR 206n).

10. In fact, in the OED under "vortex" (3657) the reference to fire precedes that of water, reflecting the etymological priority of the former to the latter: "2. An eddying or whirling mass of fire or flame. . . . 3. A whirl or swirling mass of water; a strong eddy or whirlpool."

11. Probable date of his "acme" (fortieth year). Empedocles is being considered prior to Anaxagoras in keeping with Aristotle's observation, in the *Metaphysics,* that "Anaxagoras of Clazomenae, who, though older than Empedocles, was later in his philosophical activity" (694 [984a]).

12. Translated in prose by KR.

13. The principles of Heraclitus and Empedocles, while suggesting to us a transcendent rather than an immanent mode, were nonetheless considered "material," i.e., extended spatially, by the philosophers themselves, as KR remark: "We have seen earlier . . . how gradual was the advance toward the apprehension of the abstract. Empedocles here takes another step in that direction. We shall see . . . how Anaxagoras takes yet another. But it was not until Plato elaborated his theory of Ideas that the goal was eventually reached" (330).

14. References to specific lines are to those of the poem in the original Greek.

15. Aristotle, *De Caelo,* trans. J. L. Stocks, *The Basic Works of Aristotle,* 431–32 [295a].

16. W. B. Yeats, *A Vision* (New York: Collier, 1969), 68. Based on "some fifty copy-books of automatic script, and of a much smaller number of books recording what had come in sleep" (17–18), compiled during the period of his wife's experiences of automatic writing and speaking (1917–20). Yeats's reference above to "my instructors" is to the "communicators" who spoke to him through his wife. (Letter labels have been added to one of the figures presented by Yeats to facilitate analysis.)

17. Mentioned in KR fragments 423 and 424 (326–28).

18. The degree to which νοῦς is "material" is a subtle issue. A comparison of FD's and KR's translations of a phrase from the same fragment (FD 12 and KR 503) as, respectively, "[Mind] is mixed with no Thing" and "[Mind] is mixed with nothing" shows that, in the former, the emphasis is on the lack of "thingness," whereas, in the latter, "nothing" simply indicates the absence of anything else and not necessarily of materiality itself. In the end, KR seem to compromise the issue of immanence and transcendence:

> Anaxagoras in fact is striving, as had several of his predecessors, to imagine and describe a truly incorporeal entity. But as with them, so still with him, the only ultimate criterion of reality is extension in space. Mind, like everything else, is corporeal, and owes its power partly to its fitness, partly to the fact that it alone, though present in the mixture, yet remains unmixed. (374)

19. Specifically, the cosmic philosophy of Leucippus (fl. c. 440–35 B.C.) and Democritus (fl. after c. 430 B.C.), although Epicurus and Lucretius also develop atomist doctrines, as we shall see. Because Leucippus and Democritus are practically impossible to distinguish, KR's practice of treating them together will be adopted here.

20. The issue is further complicated by Aristotle's observation that "Democritus . . . calls space by these names—'the void', 'nothing', and 'the infinite' (KR 407, fragment 555). If "infinite" and "void" are synonyms, why do they denote different loci in fragment 562?

Chapter 3

1. In *Plato's Cosmology: The Timaeus of Plato,* trans. with a running commentary by F. M. Cornford (Indianapolis: Bobbs-Merrill, 1975).

2. Plato, *The Republic,* trans. Paul Shorey (Cambridge: Harvard University Press, 1956), II, 493–95 [Stephanus II: 614C–E].

3. Plato, *The Republic of Plato,* trans. F. M. Cornford (London: Oxford University Press, 1945), 350. The author's practice of italicizing his commentary has not been duplicated here.

4. A case in point is the parallel sweeping view of the cosmos at the end of Cicero's *De Re Publica* (*On the Commonwealth*), trans. G. H. Sabine and S. B. Smith (Indianapolis: Bobbs-Merrill, 1929). In Scipio's dream, as in Er's vision, a concentrically spherical universe is beheld by the protagonist. But the Roman author offers no concrete image, relying rather on the abstract account of the system given by Scipio's father Paulus. As a matter of fact, when the latter asks his son, "Do you not perceive the heavenly spaces into which you have come?" (260), it is evident that an immediate, tangible object is lacking, and just what "spectacle" (262) Scipio sees remains ambiguous. If outside the spheres, as seems to be the case, they would only be able to see the outer sphere, if inside, only the adjacent spheres; Cicero's text fails to resolve the dilemma. By contrast, Plato's hemispherical whorl is, on the one hand, part of a common device of everyday activity, a spinner's spindle, immediately recognizable and, on the other hand, a graphic analogous embodiment of the structure of the cosmos as he envisions it.

5. After Cornford's description (PRC 353–54). In Cicero's "Stoic" cosmos, the older Greek order of the planets is revised in accordance with second century B.C. astronomy. Hence, the sequence from the moon to Mars is "moon-Mercury-Venus-sun-Mars" instead of Plato's "moon-sun-Venus-Mercury-Mars."

6. For Cicero, the earth holds the central position in a series of nine concentric spheres (260–62). Milton's allusion, ostensibly to Plato's Great Whorl in "Arcades," is to "the nine infolded Spheres" (*Complete Poems and Major Prose,* ed. M. Y. Hughes [Indianapolis: Odyssey Press, 1978], 79).

7. John Freccero discusses the relevance of this pattern to Dante in "Dante's Pilgrim in a Gyre," *PMLA* 76 (1961): 172–73; and to Donne in "Donne's 'Valediction Forbidding Mourning,'" *ELH* 30 (1963): 341–45.

8. Cornford citing Martin (PTC 131). Cicero, in "Scipio's Dream," also maintains that the earth at "the center of the universe . . . does not move" (262), although he does not explain how this is possible mechanically.

9. The definitions above in parentheses are from PTJ 34 [52A–B].

10. *Harper's Dictionary of Classical Literature and Antiquities,* ed. H. T. Peck (New York: American Book Co., 1923), 1051.

11. In "Scipio's Dream," the notion of "the music of the spheres" is similarly portrayed—seven notes to the scale, though, instead of Plato's eight—with the additional observation that continuous exposure to the strains "has filled and deafened man's ears" (263) to them. While sound waves might have been transmissible in Platonic space, a material medium, the void, we now know, exists, and the spheres revolve in stark, awesome silence.

Chapter 4

1. The dates generally agreed upon by several sources consulted are c. 342/1–270 B.C.

2. The "Letter to Pythocles" is now believed to be the work of a follower of Epicurus writing soon after his death.

3. Epicurus, *Epicurus: The Extant Remains,* trans. Cyril Bailey (Oxford: Clarendon Press, 1926), 61.

4. KR 413, fragment 567.

5. Cyril Bailey, *The Greek Atomists and Epicurus: A Study* (Oxford: Clarendon Press, 1928), 141.

6. BG 365. Serres, too, affirms the existence of a "δίνη, *dinè*, δῖνος, *dinos* . . . in Epicurus" (13).

7. R. M. Geer's note on the previously quoted passage of Epicurus (in his translation, *Letters, Principal Doctrines, and Vatican Sayings* [Indianapolis: Bobbs-Merrill, 1964], 84) is apropos:

> Our senses, of course, can give us no evidence in regard to the formation of the heavenly bodies, but they do tell us of earthly phenomena that may suggest the manner of their formation. For example, place water and a little sand in a bowl and give it a rotary motion setting up an eddy in the water. The sand will gather in the center and the water (the lighter of the two) will rise on the sides, the cross section of the whole approximating that of the lower part of a sphere. If still lighter elements like fire and air were in the whirl, we might suppose that they would rise still higher, continuing the curve until they met at the top, the whole forming a hollow sphere. Going one step further, we can imagine new eddies set up by the lighter elements in the upper part of the sphere with the fire concentrated in the center of each eddy. Here we have a small model of the creation of the world, based on the evidence of the senses or at least not contradicted by them (!)."

8. 99/96–55 B.C.

9. Lucretius, *The Way Things Are* [*De Rerum Natura*], trans. Rolfe Humphries (Bloomington: Indiana University Press, 1969), 50. All quotations from the poem in English are from this translation.

10. Line numbers in parentheses refer to the original Latin included in the following edition: Titi Lucreti Cari, *De Rerum Natura* (Libri Sex), ed. and trans. Cyril Bailey (Oxford: Clarendon Press, 1949). Bailey's prose translation of this passage reads: "nor does it mingle its liquid body with the boisterous breezes of air; it suffers all our air below to be churned by headstrong hurricanes, it suffers it to brawl with shifting storms, but itself bears on its fires as it glides in changeless advance" (457–59). "Boisterous" is perhaps a somewhat misleading translation of *turbantibus*.

11. Disorientation and confusion are the desired results in the children's game of blindman's bluff, when the person who is "it" is blindfolded and spun vigorously about. Another whirling bodily movement is the dance executed by the Sufis:

> Embodying the celestial movements of a planet around its own axis and around the sun, the Dervish, by his whirling, realizes the spiralling of the universe into being. The contraction of spirit into matter takes place around the still axis of his own heart. His right (active) hand receives the manifestation of the One and his left turns it into the earth; his spirit, like the alternate breath of the cosmos, milled free of its illusory existence, expands and spirals to its Divine Source. (PM, plate 60)

In Sufi dancing, the aims of the movement are mimetic and ritualistic, although the sense of disorientation may entrance the dancers into the nonillusory spiritual state they seek. Finally, two additional facts should be noted: (1) the passage mirrors the "divine right-handedness" and "earthly left-handedness" of previous symbolic taxonomies (e.g., Plato's); and (2) the cone-shaped robes of the dancers, spiraling out from the waist, visually enhance the symbolism of their vortical whirls (see Purce's photograph).

12. Serres also argues that Archimedes (c. 287–212 B.C.), in his treatises on spirals, cones, hydrostatics, and other turbulent configurations, "mathematized" Lucretius's Epicurean theory sufficiently to justify "the birth of physics in Lucretius." Simply stated, "The thing is in Lucretius and the theorem in Archimedes" (74). That Lucretius is more concerned with presenting the thing, not just the pure concept, which is Archimedes' intention, reveals the essentially aesthetic (sensuous) approach of the former, as opposed to the latter's noetic (scientific) method. My focus here upon aesthetic factors precludes anything but a cursory treatment of the scientific inquiries into turbulence by such thinkers as Aristotle and Archimedes.

13. Πρηστήρ, as Bailey explains in his notes to Lucretius, is the Greek term for "waterspout" and "implies the presence of fire," although the Roman poet "concentrates on wind and cloud" (LR 1618). Curiously, it was with the image of the "prester" in Heraclitus that our study of the vortex, in the ancient classical period, commenced; now it nears conclusion with the same image, one that Lucretius treats at length and charges with a greater concentration of vortical words than he does any other description in the poem.

14. The Latin verbs are *declinare* (l. 221) and *inclinare* (l. 243).

15. *Nec plus quam minimum* (l. 244).

16. *Nec regione loci certa, nec tempore certo* (l. 293).

17. *Exiguum clinamen* (l. 293).

18. According to Serres, the word *foedus* of Lucretius's *foedera naturae* connotes "the pact after the war" (148), representing a compromise between antagonistic natural forces. Plato's system requires a similar appeasement of factions through suasion.

Chapter 5

1. Kenneth Clark, *The Drawings by Sandro Botticelli for Dante's* Divine Comedy (New York: Harper and Row, 1976), 28–29.

2. "Turn, turning; circulation, rotation, revolution (of wheel, etc.); . . . circle, circuit" in *Cassell's Italian Dictionary,* compiled by P. Rebora (New York: Macmillan, 1979), 225.

3. "To turn, to rotate, to turn round, to spin round, . . . to circulate" (CI 225)).

4. Dante Alighieri, *Inferno, The Divine Comedy,* trans. with a commentary by Charles S. Singleton (Princeton: Princeton University Press, 1980). All quotations in the original Italian are from this edition and are identified, as needed, by cantica name, canto (roman numeral), and line (arabic numeral).

5. References to Singleton's prose translation are identified, as needed, by volume and part (first two arabic numerals), canto (roman numeral), and page (last arabic numeral).

6. Dante Alighieri, *Hell, The Comedy of Dante Alighieri,* trans. Dorothy L. Sayers (Harmondsworth: Penguin, 1974). Quotations from Sayers's verse translation and commentary are identified, as needed, by cantica (first arabic numeral), canto (roman numeral), and page (last arabic numeral).

7. Saint Augustine, *Confessions,* trans. R. S. Pine-Coffin (Harmondsworth: Penguin, 1973), 39.

8. The original vortical vocabulary has been interpolated, for convenience, from the facing Italian text.

9. Cf. the portraits in, for example, Sophocles' *Philoctetes* and *The Trojan Women* of Euripides.

10. Translator's capitalization.

11. "Helical" due to the progressive encirclings, indicated, for example, by Dante's reference to Virgil: " 'O supreme virtue, . . . who lead me round [*mi volvi*] as you will through the impious circles' " (DSI X:99). "Spiral" because the journey is traced on the inside of a cone, from periphery to center.

12. In Freccero's view, evidence in the Geryon episode suggests that "the descent into hell is accomplished by a clockwise spiral," although, he adds in a note, the matter is disputed (168).

13. Such is the case in Canto III, where the leap of some of the damned into hell is compared, as we have seen, to falling autumn leaves, but also, in the case of others, to the "stoop" of presumably a falcon to the falconer's command—*come augel per suo richiamo* (III:117)—although the poet does not specify the type of *augel.* In Canto XXII, the swooping down of a bat-winged demon upon a barrator roiling in pitch is compared to a falcon's dive, but the emphasis is upon the vexation of a near miss.

14. Purce notes that "The ziggurat, according to ancient Babylonian tradition, had seven windings

for the seven planets or celestial orbs" (67). The seven cornices of Dante's mountain would seem at first to correspond neatly to the ancient ziggurat, although they illustrate the "seven deadly sins" of Christian theology, and Mount Purgatory actually comprises ten main levels, as Sayers points out, when ante-purgatory, i.e., the two terraces of the excommunicate and the late repentents, and the earthly paradise are added (DSA 2:62–63). The schemes of hell and paradise are similarly decimal. The nine circles and "vestibule" of the former make ten main divisions (DSA 1:138–39), while nine revolving heavens plus the eternal empyrean of the latter account for its ten levels (DSA 3:401).

15. There is no more compelling example of the attempt, symbolically, to unite heaven and earth by fusing the circle and the square. For photographs and information since its recent restoration, see W. Brown Morton III and Dean Conger, "Indonesia Rescues Ancient Borobudur," *National Geographic Magazine,* 163:1, January 1983, 126–42.

16. Again in purgatory, as in hell, Dante's course is actually a zigzag, which abbreviates the spiro-helical pattern and thereby renders the ascent feasible in the space of thirty-three cantos (see Sayers's diagram of this upward, rightward meander [340]). Nevertheless, the revolving centripetal path from the periphery of the cone to its center that every purged soul presumably accomplishes is implicitly spiro-helical.

17. Freccero's phrase in his course on the *Divine Comedy,* Stanford University, Summer 1981.

18. Of the dictionaries consulted, only *The Cambridge Italian Dictionary,* ed. Barbara Reynolds (Cambridge: Cambridge University Press, 1962) listed "sphere" as a possible translation of *giro* (I:341–42).

19. A sphere is a three-dimensional circle. Even though the poet frequently focuses upon a circular synechdoche, such as the previously mentioned "wheel" of planetary orbits, the implied full symbol is nonetheless spherical.

20. The order of planets in Dante's Ptolemaic universe reflects the Ciceronian "correction" of Plato's design (i.e., moon-Mercury-Venus-sun-Mars-Jupiter-Saturn-Fixed Stars) with the direction of numbers reversed—the moon is number one, not eight—and the primum mobile and empyrean added as ninth and tenth levels.

21. "Imagine, too, the bell-mouth of the horn
Whose point springs from the axle's tip on high
Round which the Primum Mobile is borne."
(DSA 3:XIII:169)

22. An illustration of this phenomenon can be found in George Abell, *Realm of the Universe* (New York: Holt, Rinehart and Winston, 1976), 11.

23. Cf. "and each was moving more slowly according as it was in number more distant from the unit" (DSI XXVIII:315).

24. Cf. Plato's assigning of a siren to each concentric circle of the Great Whorl, who each intone a note of the scale, although it is the Fates who impart motion.

25. Barbara Reynolds's note.

26. Taken together, moreover, the two-tiered "cups" of the mirror image, joined at the rims, display the layered spheroidal form of a beehive.

Chapter 6

1. An earlier work, *Le Monde, ou traité de la lumière* (composed between 1629 and 1633), first introduced the vortices, as well as an explanation of gravity by the action of "subtle matter," but was suppressed by the Catholic author himself because of the condemnation of Galileo and thus appeared only posthumously. By 1672, the popularity of Descartes's theories is evident in Molière's play of that

year, *Les Femmes sçavantes,* which parodies the learned ladies' passion for science and metaphysics, including the vogue of Cartesianism:

Armande.
Epicure me plaist, et ses Dogmes sont forts.
Bélise.
Je m'accommode assez bien pour moy des petits Corps;
Mais le Vuide à souffrir me semble difficile,
Et je gouste bien mieux la matière subtile.
Trissotin.
Descartes, pour l'Ayman, donne fort dans mon sens.
Armande.
J'aime ses tourbillons.
Philaminte.
Moy ses Mondes tombans.

Armande: Epicurus pleases me, and his dogmas are strong.
Bélise: As for me, I rather like the atomic system, but I think a vacuum is hard to accept, and I relish much more the subtle matter.
Trissotin: Descartes, for magnetism, agrees with my opinion.
Armande: I love his vortexes.
Philaminte: I, his falling worlds.

(Molière, *Les Femmes sçavantes,* ed. René Bray [Paris: Les Belles Lettres, 1952], III:ii. Translation of *The Learned Ladies* by Renée Waldinger [Woodbury, N.Y.: Barron's, 1957].)

2. E. J. Aiton, *The Vortex Theory of Planetary Motions* (London: MacDonald, 1972).

3. Henri Gouhier, *Les Premières pensées de Descartes* (Paris: Vrin, 1958), 11.

4. At the request of Maxime Leroy, who translates the brief "consultation" in *Descartes, le philosophe au masque* (Paris: Rieder, 1929), I:89–90.

5. *Œuvres philosophiques,* ed. Ferdinand Alquié (Paris: Garnier, 1963), I:52–63.

6. (Vrin, 1966), X:179–88.

7. Adrien Baillet, *La Vie de Monsieur Des-cartes* (Paris: Daniel Horthemels, 1691), Part I, Book 2, Chapter 1, 80–86. Quotations from this work will retain the unabridged punctuation and spellings, which are often at odds with modern usage. Translations accompanying quoted passages rely primarily on Richard Kennington's in "Descartes' 'Olympica,'" *Social Research* 28 (July 1961): 171–204, although I have often modified them, sometimes preferring my own renderings or those of Gregor Sebba, *The Dream of Descartes,* ed. Richard A. Watson (Carbondale: Southern Illinois University Press, 1987).

8. One such passage, in which "Baillet would cease to follow the Cartesian narrative of the dream in order to comment upon the meaning accorded by Descartes," is identified by Alquié in a note (57n). Sebba speculates about many others. Even if Baillet is commenting or elaborating and not just paraphrasing—an unproven and, in the absence of the actual Cartesian text, unprovable hypothesis—the essential points are still presumably those of the philosopher.

9. It is true that between the first and the second there was an interval of two hours without sleep, that between the second and the third he was awake for a time; he was thus able to reconstruct each dream, and when, after the last, he wants to give himself a report of the entire night, his imagination recovers the memory of the reconstructions already made for the second and the first. (33)

But then Gouhier adds, "This admitted, it remains that Descartes recounts the first and the second dream after having had the third," even though the concession just made does, in my view, qualify the *tout* of his conclusion that "l'interprétation précède tout rappel." The two hours spent thinking were certainly sufficient to set the dream elements firmly in Descartes's mind before the second and third dreams occurred, and he may even have written them down before going back to sleep. Nothing in Baillet's transcription precludes such a possibility, and Sebba, in his analysis, practically assumes the existence of "hasty notes [taken] during the night" (3).

10. Made clear by his remarks concerning the significance of the dream even as he finishes recounting it (see paragraph at the top of BV 82).

11. Gouhier's study has, in effect, the first and second subordinate to the third.

12. Georges Poulet, *Études sur le temps humain* (Edinburgh: University Press, 1949), 62.

13. Cf. Baillet:

> Car il ne croioit pas qu'on dût s'étonner si fort de voir que les Poëtes, même ceux qui ne font que niaiser, fussent pleins de sentences plus graves, plus sensées, & mieux exprimées que celles qui se trouvent dans les écrits des Philosophes.
>
> For he did not believe one should be so greatly astonished to see that the poets, even those who say only foolish things, were full of maxims more serious, more sensible, and better expressed than those found in the writings of the philosophers. (84)

Poems, like dreams, express themselves in images and are not altogether subject to the rules of reason and logic that govern philosophy. Descartes clearly recognizes a privileged access to wisdom in poetic symbolism:

> La force de l'Imagination . . . fait sortir les semences de la sagesse (qui se trouvent dans l'esprit de tous les hommes comme les étincelles de feu dans les cailloux) avec beaucoup plus de facilité & beaucoup plus de brillant même, que ne peut faire la Raison dans les Philosophes.
>
> The force of imagination . . . thrusts out the seeds of wisdom (which are found in the mind of all men, like fiery sparks in pebbles) with much more facility, and even much more brilliance, than reason can do in the philosophers. (BV 84)

It is nonetheless remarkable to discover a philosopher and a confirmed rationalist like Descartes celebrating the preeminence of poetic insight. Gouhier even goes so far as to consider Descartes himself a poet: "It is the poet who opens the notebook of his reveries. For he is a poet" (23). Kennington, on the other hand, who seems to claim to be the first "historian" to offer "a detailed interpretation of the *Olympica*" (172), feels the need "to overcome the peculiar literary character" of the text and notes, condescendingly, that its "literary status . . . by no means precludes detailed study [!]" (174). He also devotes a significant portion of his discussion to devaluing Descartes's praise of poetry as merely a "sleeping interpretation" (189) that, once awake, he revises in a manner that conforms more to the negative view of poetry expressed in later works. I find the apparent conflict in Descartes's views more plausible than Kennington's extremely intricate and tenuous attempt to resolve it.

14. Descartes's remark moments later that "Par les Poëtes rassemblez dans le Recueil il entendoit la Révélation & l'Enthousiasme" ["By the poets gathered in the collection he understood revelation and enthusiasm"] does not mean that his previously mistaken "sleeping interpretation" of the significance of poetry has been "drastically revised in the waking view" (185), as Kennington claims. When poetry joins together philosophy and wisdom, it produces the kind of revelation and enthusiasm, the text goes on to

say, "dont il ne desespéroit pas de se voir favorisé" ["with which he did not despair to see himself favored"]. The latter are not attributes offered in contradistinction to the former by a now awake and alert analyst but, rather, concomitant effects of poetic inspiration and insight that he hopes will personally benefit him.

15. A marginal notation that reads *a malo spiritu ad Templum propellebar* is the basis for this conclusion, although it cannot be incontrovertibly proven. See Alquié's note (58n). Further evidence is the phrase *ventus spiritum significat,* a secondary reference to the *Olympica* in *Cogitationes Privatae* (DO, X, 218). Gérard Simon, "Descartes, le rêve et la philosophie au XVIIe siècle," *Revue des sciences humaines* 82:211 (July–September 1988), refers to "the evil Spirit—*malus Spiritus* that Baillet attenuates to *mauvais Génie*" (137). In the case of the later reference to "le Génie qui excitoit en luy l'enthousiasme dont il se sentoit le cerveau échauffé depuis quelques jours [et qui] luy avoit prédit ces songes avant que de se mettre au lit" ["the Spirit who excited in him the enthusiasm with which he had felt his brain heated for several days (and who) had predicted to him these dreams before going to bed"] (BV 85), a spirit of unknown origin and unclear moral typing, we cannot be sure of the original Latin word employed by Descartes.

16. Here, I have independently come to the same conclusion as Kennington, who observes that "none of the superhuman beings or agencies occur in dreams or sleep, but only in Descartes's waking interpretations" (175).

17. To explain the "bruit aigu & éclatant" ["sharp burst of noise"] (82), Sebba, in his nonsymbolic reconstruction of the episode, surmises that "A floorboard cracked suddenly, or there was a sound like the crack of a whip outside . . . magnified by the hypertense ear of the dreamer" (21). As for the "étincelles de feu répanduës par la chambre" ["sparks of fire scattered about the room"], he elaborates a physiological explanation involving "a muscular shock that acts like a blow on the optical nerve."

18. Sebba discovers in the motions of the first dream a physical preenactment of "the precise *itinéraire* of the *First Meditation* of Descartes" (see 49–50).

19. Poulet associates left with the "unconscious" and right with the "conscious" (67).

20. This strange and totally anomalous figure has elicited many diverse comments. Descartes's improbable interpretation of it as signifying "les charmes de la solitude, mais présentez par des sollicitations purement humaines" ["the charms of solitude, but presented by purely human solicitations"] (BV 85) is accepted at face value by virtually none of the critics (and is even derided by some); it seems, in fact, to prove the point that the dreamer is not necessarily a very reliable interpreter of his own dream. If, as Poulet argues, the melon represents the young thinker's repressed sensuality, we might expect his conscious evaluation of such an embarrassing intrusion of the unconscious to obscure all the more its true significance. The round, ripe, fleshy melon—cf. Prufrock's query, "Do I dare to eat a peach?"—constitutes, according to Poulet, "a feminine representation in erotic dreams" (74), and it is amplified by sensually charged words (*les charmes de la solitude, sollicitations*) and, I would add, by other references in the dream, such as his fear that "quelque mauvais génie . . . l'auroit voulu séduire."

Sebba recognizes the sexual symbolism but also relates it, albeit quite speculatively, to "a M. Chauveau whom [Descartes] knew at La Flèche and who came from—Melun. . . . A memory arose, and was quickly defused by a verbal pun: *Melun* becomes *melon*" (14). For Marie-Louise von Franz, "The Dream of Descartes," in *Timeless Documents of the Soul* (Evanston, Il.: Northwestern University Press, 1968), the melon not only connotes "the feminine principle" but is, among other things, an important Manichaean symbol with a "deeper meaning" in the dream, namely, "the problem of the reality of evil," which Descartes "overlooks" (116).

What I find most significant about the melon is its cryptic appearance gratuitously and arbitrarily "out of nowhere," since it is just this sort of inconsistency and non sequitur that confirms the authenticity of the dream and discredits the possibility, which has been raised and which Gouhier addresses at length (38 ff.), that it is a pure fiction. The presence of this anomaly (a detail Descartes would never rationally think to include) also suggests to me that the text we now have parallels quite closely the original dream experience and that it has not been significantly altered or embellished with the aim of producing a carefully constructed rhetorical piece like Cicero's "Dream of Scipio." In this respect, Kennington's supposition that "the work's every feature, dream as well as interpretation, is consciously intended" and that the *Olympica* is thus "a deliberate, 'poetic' construction" (185) is vitiated by the melon and other such unconscious, irrational dream elements.

21. See Martin Luther, *The Bondage of the Will,* in *Martin Luther: Selections from His Writings,* ed. John Dillenberger (Garden City, N.Y.: Anchor Books, 1961), 194–99.

22. Although Lewis S. Feuer, I recently discovered, makes the same connection in "The Dreams of Descartes," *The American Imago* 20:1 (Spring 1963): 8, he completely ignores Descartes's own interpretation and the problems it poses.

23. Françoise Meltzer, "Descartes' Dreams and Freud's Failure, or The Politics of Originality," *The Trial(s) of Psychoanalysis,* ed. Françoise Meltzer (Chicago: Chicago University Press, 1988), whose primary influence appears to have been Feuer, reaffirms his linkage of the whirlwind to the biblical image of God but also accepts Descartes's interpretation of it as "an evil genius" (86). She echoes, in this respect, Kennington, who thinks that " 'the evil spirit,' 'God,' and 'the Spirit of God' . . . are in fact one" (199). This view produces, in effect, an image of the deity that is more Manichaean than Christian and only confuses the situation. In my reading, the *mauvais génie* is not an *evil* spirit, as Descartes mistakenly thinks, but the spirit of God personified as the Old Testament's angry Yahweh in the whirlwind.

24. Apropos of which, Gouhier remarks, "Millet translated: 'things of a more spiritual and purely intelligible order'; Gaston Milhaud justly specifies 'things from on high, things divine or celestial' " (42–43). Curiously, although the Olympian allusion is a celestial one, it is decidedly non-Christian (an additional example of Descartes's having "blurred" the role of God only to exacerbate the anxiety of his *syndérèse?*). And given the generally "literary" importance and popularity of classical mythology during the Renaissance, as championed, for example, by the Pléiade poets, the title *Olympica* might actually be meant to underscore the poetic character of the experience, as discussed earlier.

25. See Heidegger's description of *Verfallen* as involving both a "downward plunge" (*Absturz*) and a "turbulent whirl" (*Wirbel*) in *Being and Time* (trans. J. Macquarrie and E. Robinson [New York: Harper and Row, 1962], 223); and Sartre's comparison of the anxiety of existential liberty to the sensation of "vertigo on the brink of the precipice" in *L'Être et le néant* (Paris: Gallimard, 1943), 64–68.

26. The psychoanalytical studies of the "trois songes" by Freud, Stephen Schönberger ("A Dream of Descartes: Reflections on the Unconscious Determinants of the Sciences," *The International Journal of Psycho-Analysis* 20 [January 1939]: 43–57) and Iago Galston ("Descartes and Modern Psychiatric Thought," *Isis* 35 [Spring 1944]: 118–28) really do not offer very much, as far as insights into the dreams and the dreamer are concerned, beyond the designation of stock sexual symbols and speculations concerning masturbation, latent homosexuality, and paranoia.

27. An approach that subsequent critics have continued unquestioningly to imitate, even relatively recent ones like Kennington, who concludes that the *Olympica* is "more 'rationalist' than the traditional interpretation of Descartes" (203), and Sebba, who states axiomatically that "The sequence develops with iron logic" (7).

28. For Feuer, the melon represents the "forbidden fruit" of this paradigmatic tree (11).

29. Jack Vrooman, *René Descartes, A Biography* (New York: Putnam's, 1970), thus entitles his chapter on the episode.

Chapter 7

1. Northrop Frye, *Fearful Symmetry* (Boston: Beacon Press, 1962), 7.

2. Neither the author nor his critics seem appreciably to distinguish the vortex from the spiral

(whether two- or three-dimensional). My primary interest is in the former, a dynamic principle, although it is apt to be represented, in the illustrations at least, by the convoluted form of the latter.

3. W.J.T. Mitchell, *Blake's Composite Art* (Princeton: Princeton University Press, 1978), 69.

4. Numbering added.

5. William Blake, *The Four Zoas, The Complete Poetry and Prose of William Blake,* ed. David V. Erdman (New York: Anchor Press, 1988). I have adopted the system of page and line reference used in this edition.

6. Kathleen Raine, *Blake and Tradition* (Princeton: Princeton University Press, 1968).

7. Martin K. Nurmi, *William Blake* (London: Hutchinson University Library, 1975).

8. Donald D. Ault, *Visionary Physics: Blake's Response to Newton* (Chicago: University of Chicago Press, 1974).

9. For examples, see AV 45 (fig. 4) and 50 (fig. 6).

10. S. Foster Damon, *A Blake Dictionary* (Providence: Brown University Press, 1965), states that "*The Four Zoas* is Blake's rewriting of *Paradise Lost*" (275), according to which the creator of the material world was not God the Son, as Milton has it, but rather Blake's diabolical Urizen.

11. S. Foster Damon, *Blake's Job* (Providence: Brown University Press, 1966), 32–33.

12. Albert S. Roe, *Blake's Illustrations to the Divine Comedy* (Princeton: Princeton University Press, 1953).

13. John E. Grant, "Visions in Vala: A Consideration of Some Pictures in the Manuscript," *Blake's Sublime Allegory,* ed. S. Curran and J. A. Wittreich, Jr. (Madison: University of Wisconsin Press, 1973), 184.

14. Blake, *Milton,* in *The Complete Poetry and Prose of William Blake.*

15. Ronald L. Grimes, "Time and Space in Blake's Major Prophecies," *Blake's Sublime Allegory,* 80.

16. Harold Bloom, *Blake's Apocalypse* (Garden City, N.Y.: Doubleday, 1963), 325, concurring with the views of Frye and Hazard Adams.

17. Brian Wilkie and Mary Lynn Johnson, *Blake's Four Zoas: The Design of a Dream* (Cambridge: Harvard University Press, 1978), 132.

18. Inscribed at the base of the title page.

19. See Darrell Figgis, *The Paintings of William Blake* (London: Ernest Benn, 1925), pl. 80.

20. The fourfold creatures (each with four faces, four wings) of the biblical text are repeatedly associated by Blake with his Four Zoas and recur in a variety of contexts throughout his works.

21. DSA XXIX: 305. Roe, citing others, agrees (166).

22. Anne K. Mellor, *Blake's Human Form Divine* (Berkeley: University of California Press, 1974), 261.

23. See, for example, Dionysius the Areopagite, *On Divine Names, The Works of Dionysius the Areopagite,* trans. Rev. John Parker (Merrick, N.Y.: Richwood, 1976), who states that "a soul is moved spirally, in so far as it is illuminated, as to the divine kinds of knowledge, in a manner proper to itself, not intuitively and at once, but logically and discursively; and, as it were, by mingled and relative operations" (I:IX [42–43]).

24. Cf. Dionysius's description of three types of angelic movement:

> Now, the divine minds [angels] are said to be moved circularly indeed, by being united to the illuminations of the Beautiful and Good, without beginning and without end; but in a direct line, whenever they advance to the succour of a subordinate, by accomplishing all things directly; but spirally, because even in providing for the more indigent, they remain fixedly, in identity, around the good and beautiful Cause of their identity, ceaselessly dancing around (I:VIII [42]).

Chapter 8

1. He is strongly linked to it, though, through Baudelaire, widely considered the father of French symbolism. Poe, certainly as much as anyone else, was for Baudelaire a poetic and aesthetic mentor.

2. Charles Baudelaire, "Le Gouffre" ["The Abyss"], *Œuvres complètes,* ed. Marcel A. Ruff (Paris: Seuil, 1970), 88.

3. Richard Wilbur, "The House of Poe" (Library of Congress Anniversary Lecture, 4 May 1959), *The Recognition of Poe: Selected Criticism Since 1829,* ed. Eric W. Carlson (Ann Arbor: University of Michigan Press, 1970), 257.

4. Edgar A. Poe, "MS. Found in a Bottle," *The Complete Works of Edgar Allan Poe,* ed. James A. Harrison (New York: AMS Press, 1965), II:1. References to this work identify volume and page numbers.

5. David Halliburton, *Edgar Allan Poe: A Phenomenological View* (Princeton: Princeton University Press, 1973), 248.

6. Poe's style is similarly polarized, according to Donald B. Stauffer, "The Two Styles of Poe's 'MS. Found in a Bottle,'" *Style* I:2 (Spring 1967), who asserts that "the tale is written in two quite different styles, to which we might apply the terms 'plausible' and 'arabesque'; and the interweaving of these two styles, one of which predominates at the beginning and the other at the end, gives the story its texture of mixed fact and fantasy" (108).

7. The link with Dante has been recognized by others, as have similarities with Coleridge's "Ancient Mariner." See Stephen L. Mooney, "Poe's Gothic Waste Land" (*Sewanee Review,* January–March 1962) in *The Recognition of Edgar Allan Poe,* 281–82.

8. Although the whirlpool is not explicitly mentioned in this episode, I think it is implied by the combination of "dizziness" (caused by a spinning motion) and "descent."

9. A quotation from Philippe Quinault's *Atys* (1676) that reads:

> Qui n'a plus qu'un moment à vivre
> N'a plus rien à dissimuler.
>
> He who has no more than a moment to live
> Has nothing more to dissimulate.

10. Harold Beaver, "Doodling America: Poe's 'MS. Found in a Bottle,'" *A Centre of Excellence: Essays Presented to Seymour Betsky,* ed. Robert Druce (Amsterdam: Rodopi, 1987), suggests that the narrator "was lost in a whole visual and oral semiology without a 'key'. At the crux of his tale, that is, he had slipped into Lacan's matrix of linguistic consciousness, that unvoiced source of speech where the ties between signifier (S) and signified (s) may become unattached and broken. His signified, that is, were devoid of signifiers (a 'feeling, for which I have no name') and signifiers devoid of signified (the captain's 'language which I could not understand')" (25).

11. Stephen Peithman, *The Annotated Tales of Edgar Allan Poe* (Garden City, N.Y.: Doubleday, 1981), notes that "the O.E.D. cites Poe's use of the umlaut over the 'o' in 'Maelström' as the only such use" (96).

12. Which is also, paradoxically, a bottle in a manuscript, since, as Ken Frieden remarks in "Poe's Narrative Monologues," *Genius and Monologue* (Ithaca, N.Y.: Cornell University Press, 1985), rpt. in *The Tales of Poe,* ed. Harold Bloom (New York: Chelsea House, 1987), "the bottle only exists by virtue of the text 'inside' that describes its existence" (145).

13. Erich Pontoppidan, *The Natural History of Norway* (London, 1755), vol. I, sec. X. The account of Jonas Ramus that Poe quotes in the story matches nearly word for word Pontoppidan's presentation of it, but the Bishop of Bergen also reproduces and praises the "uncommon erudition and genius" of Ramus's "proofs" (with which he "can by no means agree") that "Scylla and Charybdis,

which have always been accounted to lie upon the coast of Sicily, were no other than this Moskoestrom, whither Ulysses was actually driven in the course of his wanderings" (84–87).

14. Gerard M. Sweeney, "Beauty and Truth: Poe's 'A Descent into the Maelström,'" *Poe Studies* 6 (June 1973): 22.

15. Robert Shulman, "Poe and the Powers of the Mind," *ELH* 37 (1970): 252–53, cited by Sweeney (25n).

16. In this sense, one could say, the rightward vortex "makes things right" by counteracting the brother's iniquity and by affirming the mariner's resourceful discovery of a secret of nature, dispensing thereby a kind of natural justice.

17. Marie Bonaparte, *The Life and Works of Edgar Allan Poe* (London: Imago, 1949), 352.

18. Poe, *Eureka*, XVI, 186.

Chapter 9

1. Arthur Rimbaud, Letter to Georges Izambard, 13 May 1871, *Œuvres*, ed. Suzànne Bernard (Paris: Garnier, 1960), 343–44. Translations accompanying Rimbaud's text are my own, although I have not hesitated to consult for comparison those of Wallace Fowlie (Rimbaud, *Complete Works, Selected Letters* [Chicago: University of Chicago Press, 1966]) and Louise Varèse (Rimbaud, *Illuminations and Other Prose Poems* [New York: New Directions, 1957]).

2. Letter to Paul Demeny, 15 May 1871 (R 348).

3. "H" (R 303).

4. "Parade" (R 261).

5. Tzvetan Todorov, "Une Complication de texte: les 'Illuminations,'" *Poétique* 34 (April 1978): 252.

6. Ross Chambers, "To Read Rimbaud (a) Mimesis and Symbolisation: A Question in Rimbaud Criticism," *Australian Journal of French Studies* 11:1 (January–April 1974): 55.

7. In his article on Rimbaud to which I have been referring, Todorov addresses in detail the hermeneutical problem the poet poses, distinguishing four main types of interpretation that have been applied to him. Two of these, "euhemeristic criticism," which treats the text "as a source of information on the life of the poet" (241), and "etiological criticism" by which "one examines the reasons that influenced Rimbaud to express himself in such a way" (242), are not strictly concerned with the signification of the text itself and, while interesting and perhaps even useful for a fuller appreciation of Rimbaud, are both extratextual and extrahermeneutical. The other two, on the contrary, "do indeed involve interpretation" and include "esoteric criticism" and "paradigmatic criticism." In the former, "each element of the text, or at least each problematic element, is replaced by another, drawn from any variant of universal symbolism, from psychoanalysis to alchemy" (243) and Todorov concludes that "these interpretations can never be confirmed, nor invalidated, whence their lack of interest" (243). I do not agree, however, with his dismissal of the latter and find his arguments against it predicated on specious reasoning (an issue I cannot develop here). The paradigmatic approach, while not applicable perhaps to every work, does lend itself to particular interpretive difficulties inherent in texts like Rimbaud's since, as Todorov himself points out in his own insightful definition of it,

> One proceeds here from the postulate, explicit or implicit, that the continuity is deprived of significance; that the task of the critic is to bring together elements more or less estranged in the text, in order to show their resemblance, opposition, or kinship; in a word, that the paradigm is pertinent, but not the syntagm. (243)

8. Cf. Heraclitus's and Lucretius's *prester,* studied previously. While the vortex image has creative-destructive significance for both authors, only Heraclitus's symbol evinces the oxymoronic properties of fire and water (implied in the Rimbaldian trope) that indicate intense conditions of violent opposition and conflict. "Le Bateau ivre" is widely interpreted to be a poem about the process of artistic (poetic) creation, which nearly always involves, for Rimbaud, the "destruction" of established conventions in order to arrive at "the new" or "the unknown," as stipulated in the "Lettres du voyant" ["letters of the seer"]. Consciously or not, the young poet thus taps a rich symbolic vein with these ostensibly incidental and relatively undeveloped vortical images.

9. There is also a whirling (*tournoyant*) *dance macabre* in the nightmarish fantasy, "Bal des pendus" ["Dance of the Hanged Men"] (R 48–49).

10. The ensuing syntactic analysis of "Marine," arrived at independently, parallels in certain respects Jacques Plessen's close reading in "*Marine* de Rimbaud: une analyse," *Neophilologus* 60:1 (January 1971): 22–23. My focus, however, is exclusively upon elements that contribute to turbulent "confusion."

11. Hugo Friedrich, *The Structure of Modern Poetry,* trans. Joachim Neugroschel (Evanston, Il.: Northwestern University Press, 1974), 61.

12. Jean-Pierre Giusto comes to more or less the same conclusion in *Rimbaud Créateur* (Paris: Presses Universitaires de France, 1980), 229.

13. The word "angle," connected to each phrase by the syntactically equivocal *dont,* clearly denotes the intersecting streamlines. This "mechanical" term, as well as the final "whirling motion," which seems to Plessen to be "the solution of a mechanical problem concerning compound movement" (24), are evidence, according to him, of Rimbaud's " 'scientific' preoccupations" (31n). Technical terminology of this sort, which abounds in the *Illuminations,* does, in any case, reflect the poet's stated preference for and attempt to realize "la poésie objective" (R 343).

14. Atle Kittang, *Discours et jeu: essai d'analyse des textes d'Arthur Rimbaud* (Bergen, Norway: Universitetsførlaget, 1975), 252.

15. "Vertigo" or "Impression by which a person believes that the surrounding objects and himself or herself are animated by a circular movement or by oscillations" (*Dictionnaire Robert*) is a sensation that may result from any number of causes. I am mainly interested in it as a "symptom" of whirling. Paule Lapeyre, *Le Vertige de Rimbaud: clé d'une perception poétique* (Neuchâtel: La Baconnière, 1981), seems to suggest that *vertige* is endemic to all of Rimbaud's poems and goes so far as to claim that "Rimbaud has appeared to us as though 'programmed' with vertigo in his genetic code" (21), an exaggeration that weakens the credibility of her interpretation.

16. Although the more idiomatic English translation of "la vraie vie" would be "real life," an important distinction must be made here between "real" (Rimbaud does not use the French *réelle*) and "true." The poet is clearly referring to a transcendent state of "true" being, not to existence in the immanent "real" world of things (*res*).

17. Anglicized form of the French term *alinéa* used to designate the "poetic paragraph" that is the typical unit of subdivision in a prose poem.

18. Which also occurs as a fortunate coincidence in the words *cône* ["cone"] and *abîme* ["abyss"], although in the case of the latter, the circumflex must be mentally reversed to reflect visually the word's meaning. Whether or not the poet actually intended the signifier to echo the sign's symbolism is rendered moot by the fact that it does.

19. Cf. "la fanfare tournant" ["the fanfare turning"] of "Matinée d'ivresse" ["Morning of Drunkenness"] (R 269), where the turning probably indicates the vertigo of hashish intoxication ("Ce poison," "*Assassins,*" etc.) but where the circular bells of the horns also seem to imbue the fanfare itself with circular properties as the sound passes through them. In a similar manner, the spiro-circular convolutions of the sea conches inform the sound they emit.

20. Anne Freadman, "To Read Rimbaud (b) A Reading of 'Mystique,' " *Australian Journal of French Studies,* 11:1 (January–April 1974), 77–78.

21. Letter to Paul Demeny, 15 May 1871 (R 346).

22. Shoshana Felman, "'Tu as bien fait de partir, Arthur Rimbaud': Poésie et modernité," *Littérature* 11 (October 1973), exploits the semantic depth and suggestiveness of many of Rimbaud's famous utterances, including this one, and cites possible "linguistic," "geographic," and "carnal and physiological" significations of the word *sens* (11).

Chapter 10

1. Wallace Fowlie, *Mallarmé* (Chicago: University of Chicago Press, 1970 [first published 1953]), 221. Fowlie does not mean that *Un Coup de dés* represents a personal failure for Mallarmé but that it elaborates thematically "a cosmic elemental tragedy." He very rightly situates it in the context of such "sea poems" as "Brise marine" and the octosyllabic sonnets, "Salut" and "A la nue accablante tu." Pierre Saint-Amand, "Mallarmé: Tourbillons" (*MLN* 95 [1980]), also discovers images and ideas of the later poem in earlier works and argues further that "the dance in Mallarmé is always described through the image of the vortex" (1346n), as in the following "dance/sea" fused figure of "Billet à Whistler":

> Mais une danseuse apparue
>
> Tourbillon de mousseline ou
> Fureur éparses en écumes
>
> But a dancer apparent
>
> Whirl of muslim or
> Fury scattered in foam

And both critics—indeed most critics—recognize the generation of many of the great poem's themes and symbols in *Igitur.* Although I shall have occasion to comment upon the "descent by spiral stair" of this *conte* ["tale"], I do not think the vortical figure, specifically, is adequately developed in any of the earlier works to warrant exclusive scrutiny. It is in *Un Coup de dés* alone that it bursts forth as a compelling, elaborate last symbol-synthesis of Mallarmé's poetic existential metaphysics.

2. Gardner Davies, *Vers une explication rationnelle du* Coup de Dés (Paris: Corti, 1953).

3. Approximate English translation of Mallarmé's own self-description as a "syntaxier" in a remark to Maurice Guillemot, cited by Mondor in his biography (*Vie de Mallarmé* [Paris: Gallimard, 1941–42], 506–7) and again by Scherer in *L'Expression littéraire dans l'œuvre de Mallarmé* (1947), who further asserts that "it would be necessary to give 'syntax' an extremely broad meaning" (79) and prefers Robert de Montesquiou's term *agencement* ["arrangement"] or Leo Spitzer's *Beziehungslehre* ("study of relations"). Derrida similarly calls attention to this neologism in *La Dissémination* (Paris: Seuil, 1972), "a word," he avers, "that I have never encountered elsewhere, not even in Mallarmé" (206). Todorov has more recently identified Rimbaud as a "lexical" poet "contrary to the 'syntaxer' Mallarmé" (TC 246).

4. Fowlie's remarks, in this regard, are worth noting: "A thought is a living irony, formed by successive births and finally lost in the pure absence of death. The language of a thought can have no absolute meaning because of the chance it encounters in its expression. Poetry is a game of risk, of magic and incantation. Its meaning is always hidden under the brilliance of its images and the unusualness of its analogies" (223).

5. See his Appendix, 200–206. Charles Chadwick, *Mallarmé: sa pensée dans sa poésie* (Paris: Corti, 1962), asserts that Davies "has unraveled . . . the syntactic structure of the poem" (136n) and proceeds to

offer his own reading of the narrative line, treating it, for the most part, as a veiled autobiographical account of Mallarmé's failure to accomplish his *Grand Œuvre.* Such linear decipherings, while affording some limited insights into the work, tend to understate its symbolic polyvalence and run counter to Mallarmé's statement in the *Préface* that "Tout se passe, par raccourci, en hypothèse; on évite le récit" ["Everything happens, by abridgment, hypothetically; one avoids narrative"] (*Œuvres complètes,* ed. Henri Mondor and G. Jean-Aubry [Paris: Gallimard, 1945], 455) and that the poem represents, rather, a new "genre" analogous to "la symphonie" (456). Suzanne Bernard, "Le 'Coup de dés' de Mallarmé replacé dans la perspective historique," *Revue d'histoire littéraire de la France* 51 (April–June 1951), rightly observes that "the 'music' of the poem is not therefore linear and successive, but polyphonic" (191). She, Roulet (whom she echoes), and Pierre-Olivier Walzer (*Essai sur Mallarmé* [Paris: Seghers, 1963], 236–37), among others, emphasize the "contrapuntal" nature of the poem-score, which phrase-by-phrase reconstructions of a story line (fable, allegory, myth) are unable fully to encompass.

6. Pagination follows that of R. G. Cohn's reduced version of the poem, included as a separate booklet in a rear pocket of his study, *Mallarmé's* Un Coup de Dés, *an exegesis* (New Haven: Yale French Studies, 1949). The word "page" denotes one of the eleven double pages numbered by him. Of all the editions to date, Mitsou Ronat's meticulously reconstructed folio format (Paris: Change errant/d'atelier, 1980) is clearly the one that comes closest to realizing Mallarmé's final conception, as Virginia La Charité demonstrates in *The Dynamics of Space: Mallarmé's* Un Coup de dés jamais n'abolira le hasard (Lexington, Ky.: French Forum, 1987), chapter 2. Whereas Cohn's version contains fourteen leaves, including covers, Ronat has dispensed with covers and printed the title on the front of the "UN COUP DE DÉS" double page. In this way, the poem displays "twelve" leaves and conforms to her argument that the number twelve is a sort of magical "golden number" (page 2 of her accompanying article, "Cette architecture spontanée et magique") for the poem as a whole. In the end, Ronat's interpretation of the significance of her "little discovery" seems unresolved, beyond the conclusion that Mallarmé "was always concerned with the material fabrication of his poems" (6). Moreover, her division of the poem into twenty-four single pages is unsatisfactory, since it diminishes the importance of the double page. That Mallarmé conceived of the work as a series of double pages is apparent in the way his handwritten notations on the corrected Lahure proofs (reprinted by Cohn in *Mallarmé's Masterwork: New Findings* [The Hague: Mouton, 1966]) treat double pages as single units, with crossover lines and other such markings frequently linking verso and recto "halves."

Translations of Mallarmé's text are my own, although I have at times consulted established translations for comparison, particularly those of Keith Bosley (Mallarmé, *The Poems* [Harmondsworth: Penguin, 1977]); Mary Ann Caws and Brian Coffey (Stéphane Mallarmé, *Selected Poetry and Prose,* ed. Mary Ann Caws [New York: New Directions, 1982]); and Anthony Hartley (*Mallarmé* [Harmondsworth: Penguin, 1970]).

7. The paradigm is extended further, as some critics have observed, by the culminating image of "UNE CONSTELLATION" (11), although the light and dark values are reversed, as in a photographic negative, with the dark words-stars distributed on light page-space.

8. Like Warren Ramsey, "A View of Mallarmé's Poetics," *The Romanic Review* 46:3 (October 1955), and unlike most other critics, I propose using this uncommon—and thus semantically underdetermined—word instead of "ideogram," a term that does not seem suited to the unique typographic experiment undertaken by the poet. Many typical ideograms are signs that often bear little or no resemblance to the "ideas" they represent; the keyboard on which these words are being typed, for example, includes a dollar sign, cent sign, and ampersand, three common ideograms that are ostensibly arbitrary figures. Mallarmé, on the other hand, intended the typographical disposition of each double page to illustrate, to some degree at least, important themes of the poem, or so his reference to "l'estampe" [print] traced by the letters, in a note to Gide quoted by Walzer (238), seems clearly to suggest. One might then argue that *calligramme* is the proper term—"Poem in which the lines are assembled in such a way as to represent an object" (*Dictionnaire Robert*)—but its close association with poems by Apollinaire published many years after the creation of *Un Coup de dés* renders it anachronistic when applied to Mallarmé. Apollinaire's *Calligrammes,* moreover, usually display an unmistakable correspondence between typographical form and poetic

content and utilize techniques, such as circular, backward, or vertical dispositions of letter sequences, that are alien to Mallarméan practices. Walzer, in fact, stresses the difficulty with which the illustrated objects or themes in *Un Coup de dés* are identified: "if the poem does indeed speak of a listing vessel, of a feather floating on the water and of a constellation perforating the sky, a singularly shrewd eye is needed to perceive the graphic on the page before having read the text—and even after" (236). Walzer's use of the word "graphic" (*le graphique*) appropriately denotes the print-like image the poet has attempted to graph (however suggestively) on at least some of the pages. Hence, "ideograph," which the dictionaries treat as a somewhat more generalized synonym of "ideogram," has the advantage of directly connoting the notion of a graphic design, as well as incorporating the important Mallarméan concept of the "idea."

9. Carol Barko, "The Dancer and the Becoming of Language," *Yale French Studies* 54 (1977), perceives, in this last image, "the intense 'jouissance' of Eros-Thanatos ('tourbillon d'hilarité et d'horreur')—the modifiers invoking the erotic 'rire' of a sexual death (continued on Page 8) and the horror of the moment in which the self knows absolutely its movement toward death" (184).

10. A purported "masculine" and "feminine" thematic that runs through his analysis, however, does not seem particularly evident or pertinent.

11. "... *la neutralité identique du gouffre*" ["the identical neutrality of the abyss"] (9).

12. "*plume solitaire éperdue*" (7).

13. This sentence also represents an intersection of the bifurcated title phrase and subsidiary *phrase hypothétique*—the latter depending on the former for its completion—as well as a conjunction of the roman and italicized types. The "*SI*," which appears on the previous page (8), is immediately linked, when the page is turned, to the rest of the phrase, because of its italic type and prominent size and because the highlighted words of page 9 readily complete the *SI* proposition that the reader anticipates. Nevertheless, the isolation of "*SI*" on page 8 and its unique status as the only word in 15-point bold italics also keep it syntactically open, an ambivalence reinforced by its semantic depth: "is '*SI*' (8) a conjunction, an intensive adverb or an affirmative adverb?" Malcolm Bowie wonders in "The Question of *Un Coup de dés*," *Baudelaire, Mallarmé, Valéry: New Essays in Honour of Lloyd Austin*, ed. Malcolm Bowie, Alison Fairlie, and Alison Finch (Cambridge: Cambridge University Press, 1982), 146. While it seems clear that "*SI*" is a conjunctive generator of hypothesis, might it not also anticipate the affirmation of art on the last page by proposing as a response to such negative pronouncements as the title phrase an equivocal "yes/if"?

14. Malcolm Bowie, *Mallarmé and the Art of Being Difficult* (Cambridge: Cambridge University Press, 1978), 129–30.

15. See René Alleau, *La Science des symboles* (Paris: Payot, 1976), 33.

16. A "chiasmus," rooted etymologically in the Greek letter chi (X), connotes the convergence of inverted terms of a proposition, but it also describes a principle of both intersection and circularity, since the proposition ends as it begins.

17. Note, again, the "fusion" of roman and italicized type and type sizes.

18. Cf. "ancestralement" ["ancestrally"], the powerful opening adverb (and Mallarméan neologism) on page 5.

19. In the famous letter to Verlaine of November 16, 1885, now entitled *Autobiographie*, Mallarmé proclaims, as a primary goal of his projected *Grand Œuvre* or *Livre*, "L'explication orphique de la Terre, qui est le seul devoir du poëte et le jeu littéraire par excellence" ["The orphic explanation of the Earth, which is the sole duty of the poet and the literary game par excellence"] (M 663). Certain critics (e.g., Roulet) situate *Un Coup de dés* squarely in the context of the poet's ambitious, if unfinished, grand design. Cohn emphasizes the "orphic" nature of the title phrase (9), and several of the poem's principal themes, highlighted typographically, definitely do evince an oracular, enigmatic, even mystical ambiguity. Such evidence seems to constitute some proof, then, that *Un Coup de dés* belongs to the *Grand Œuvre*, despite the objections of Svend Johansen, "Le Problème d'*Un Coup de dés*," *Orbis Litterarum* 3:1 (1945), who concludes that such a link "rests therefore on very weak arguments" and that the poem may well be "an isolated work" (288–89).

20. When the dice come to rest exhibiting "the unique Number that cannot be another," chance-becoming is, for that throw ("*un* coup de dés"), that particular event, fixed, determined, defeated, as it were, and so, transcended; but chance itself ("*le* hasard") is by no means "abolished."

21. The dizzying sensation of the whirl, moreover, is intimated by the vortical synecdoche, "*vertige*" (8), and the diving twist of the mermaid ("*torsion de sirène*"), on the same page, anticipates (prefigures) the imminent spiro-vortical plunge of the sailor/feather/poet.

22. Saint-Amand (1344n).

23. Claude Roulet, *Eléments de poétique mallarméenne d'après le poème* Un Coup de dés jamais n'abolira le hasard (Neuchâtel: Editions du Griffon, 1947), states: "In short, the work is a modern *De natura rerum*" (19). His subsequent "biblical" reading of the poem, however, strikes me as far-fetched. For a detailed critique, see Johansen (289–302).

24. Evoking also the sloping obliquity of the vortical cone.

25. "Plane désespérément" ["soars (hovers) hopelessly"] (3).

26. The spiral (helical?) descent recalls Dante's spiro-helical pilgrimage to the underworld, as well as Blake's spiro-vortical depictions of transcendence from one zoa to another. For Thomas Williams, *Mallarmé and the Language of Mysticism* (Athens: University of Georgia Press, 1970), *Igitur* portrays a "katabasis," i.e., a "descent of the soul into the nether world and death and its subsequent resurrection to a new life" (74). The incomplete, fragmented nature of the text, admittedly, makes interpretation all the more difficult, but I fail to detect the positive sense of "spiritual progress" and enlightenment that Williams claims to have found. The "spiral descent" does, in any case, prefigure the spiro-helical vortical descent of *Un Coup de dés,* although there is an important symbolic difference. One chooses to descend a stair, as one opts to commit suicide, but one is drawn willy-nilly to destruction if one "chances to fall" within the influence of a vortex.

27. Baudelaire, "Le Voyage" (124).

28. "Such as into Himself at last eternity changes him," from "Le Tombeau d'Edgar Poe" (M 70).

Appendix

1. "Spiral" (OED 2966).

2. Horace F. Judson, *The Search for Solutions* (New York: Holt, Rinehart and Winston, 1980), 36.

3. Theodore A. Cook, *The Curves of Life* (London, 1914; rpt. New York: Dover, 1979), calls this figure a "flat" spiral (24); Thompson uses the term "discoid," which he distinguishes from "turbinate" forms in his study of shellfish (812). The latter is a three-dimensional type that will be taken up presently.

4. Inversely, the moon is spiraling centrifugally away from the earth, as a comparison of ancient and contemporary "growth rhythms" of the nautilus, itself spiral-shaped, would seem to indicate. Hence, according to Kahn and Pompea, "420 million years ago the moon circled the earth once every nine days, the day had only twenty-one hours, and the moon loomed enormous in the sky—less than half its present distance from the earth" (JS 38).

5. The concentric layers in wood and on the fingertips are not technically spiralic, as center and circumference are not connected by a single, continuous path. The principle is, rather, a regular circular expansion in all directions, like the propagation of concentric circular waves in a ripple tank. This pattern is obviously related to the spiral, because a dynamic relationship between center and periphery is implied, but unlike the spiral's "direct link," the emphasis is upon the autonomy of the distinct stages, which are, in effect, discrete, self-contained circles.

6. A similar principle of energy storage and release is manifested, according to Cook, in the human heart, which in cross section is a flat spiral, and he attributes its effectiveness as a pump to the torsion of

its spiral construction. A spiral spring, like the mainspring of a wound watch, is another mechanism that exploits the tension inherent in coiling certain materials into a spiral.

7. If a continuum of adjacent circles is produced by translating a circular motion laterally across a plane, like the series of tightly connected loops formed by repeated gyrations of the arm in certain handwriting drills aimed at perfecting the letter O, one might argue that a two-dimensional helix is produced. I have never seen this term applied to such a configuration, and since it is not the form that the unqualified term "helix" readily calls to mind (which is, as we have seen, three-dimensional), it would have to be designated specifically as two-dimensional to avoid confusion.

8. See *Identity and Difference,* trans. Joan Stambaugh (New York: Harper and Row, 1969), 23–28.

9. Vis-à-vis of which Thompson offers the following insight concerning the organic spiral: "The shell-less molluscs are never spiral; the snail is spiral but not the slug. In short, it is the shell which curves the snail, and not the snail which curves the shell" (767).

10. The choice of this term was the logical consequence of the previously distinguished "spiral" and "helix," since the figure in question incorporates elements of both. I was nonetheless very interested to discover the evolution of more or less the same terminology in D'Arcy Thompson's study of spiralic and helical structures. Rather than presenting a finished, precise nomenclature, his text records the process of grappling with labels in an attempt to find the appropriate terms and definitions. In his distinction of the "turbinate" shell from the "discoid" type, he first observes that since the former "partakes, therefore, of the character of a helix, as well as of a logarithmic spiral; it may be strictly entitled a helico-spiral" (812). Later, in his discussion of wild goats' horns and of "the helicoid component of the spire" which accounts for the horns' divergence from a strict plane spiral, he calls such a form "a helical spiral in space" (880). Finally, he arrives at the term "spiral helix" (which, I must admit, I deduced independently following a similar process of reasoning) in his examination of cases of phyllotaxis, in which the circumvolutions wind about a conical stem, tapering off toward the apex (918).

11. "Gyre" (OED 1234).

12. *Van Nostrand's Scientific Encyclopedia* (Princeton: Van Nostrand, 1958), 1784.

13. J. Thewlis, ed., *Encyclopaedic Dictionary of Physics* (Oxford: Pergamon Press, 1962), VII:669. The axis is also termed the vortex "filament," defined as "the locus of the centers of circulation" (VN 1784).

14. *Grand Larousse encyclopédique* (Paris: Larousse, 1960 and 1964), X:408–9.

15. *McGraw-Hill Encyclopedia of Science and Technology* (New York: McGraw-Hill, 1977), XIV:431.

16. "Hydraulics," *Encyclopaedia Britannica,* 11th ed. (Cambridge: Cambridge University Press, 1910), XIV:45.

17. Pierre Larousse, *Grand Dictionnaire universel* (Paris: Administration du Grand Dictionnaire universel, n.d.), XV:356.

18. The term preferred by MH (431) and TE (670).

19. *La Grande Encyclopédie* (Paris: Société Anonyme de la Grande Encyclopédie, n.d.), XXXI:234.

20. Harry L. Shipman, *Black Holes, Quasars, and the Universe* (Boston: Houghton Mifflin, 1976), 65.

21. Martin Gardner, *The Ambidextrous Universe* (New York: Basic Books, 1964), 18.

22. In his "first great experiment," Pasteur demonstrated that the capacity of certain crystals to rotate polarized light to the right (dextrorotary) or to the left (levorotary) "established beyond doubt that molecules were capable of existing in enantiomorphic, mirror-image forms" (GA 108).

23. In a further complication of the problem, imagine yourself descending backward, the newel always to your left, but is the direction then leftward?

24. In specific examples, it will be necessary to indicate the important relative factors of the observer's position vis-à-vis the circuit: in a vertical helix, for example, whether one ascends or descends, and in the case of a spiral or spiral helix, whether the motion is centripetal or centrifugal.

25. Cf. the positive and negative values of such terms as "dextrous" and "sinister," derived from the Latin words for right and left.

26. They are either discoid, logarithmic spirals or spiral helices (to the degree that they translate

along an axis) and seem to display the full range of possible spiral and spiro-helical orientations, as Cook's twelfth and Thompson's thirteenth chapters vividly illustrate.

27. James D. Watson, *The Double Helix* (New York: Atheneum, 1969), 200. Viewed vertically, however, the structure ascends in a sinistral, counterclockwise direction.

28. Peter Gwynne with John Carey, "A New Type of Life," *Newsweek*, 24 December 1979, 58.

29. A lesser-known single helical caduceus exists that may be related to the staff of Aesculapius.

30. Stiskin points out a similar paradox in the Tao symbol. Even though the two magatamas appear ostensibly to be opposites—the one white, tail end down, the other black, tail end up—they are not mirror images, since both logarithmic spirals turn in the same direction (see SL 116–17).

31. *Asimov's Biographical Encyclopedia of Science and Technology* (Garden City, N.Y.: Doubleday, 1964), 239.

32. A common observable example is the spiraling of water through a drain, but, according to Gardner, "the question whether the Coriolis force is sufficiently strong to be detectable as an influence," producing oppositely handed directions of gyration for drains in the northern and southern hemispheres, is "undecided" (48) and awaits more conclusive experimental evidence.

Works Cited

Abell, George. *Realm of the Universe.* New York: Holt, Rinehart and Winston, 1976.

Aiton, E. J. *The Vortex Theory of Planetary Motions.* London: MacDonald, 1972.

Alleau, René. *La Science des symboles.* Paris: Payot, 1976.

Aristotle. *The Basic Works of Aristotle.* Ed. Richard McKeon. New York: Random House, 1941.

Asimov, Isaac. *Asimov's Biographical Encyclopedia of Science and Technology.* Garden City, N.Y.: Doubleday, 1964.

Augustine, Saint. *Confessions.* Trans. R. S. Pine-Coffin. Harmondsworth: Penguin, 1973.

Ault, Donald D. *Visionary Physics: Blake's Response to Newton.* Chicago: University of Chicago Press, 1974.

Bachelard, Gaston. *La Poétique de l'espace.* Paris: Presses Universitaires de France, 1978.

Bailey, Cyril. *The Greek Atomists and Epicurus: A Study.* Oxford: Clarendon Press, 1928.

Baillet, Adrien. *La Vie de Monsieur Des-cartes.* Paris: Daniel Horthemels, 1691. Part I, Book 2.

Barko, Carol. "The Dancer and the Becoming of Language." *Yale French Studies* 54 (1977): 173–87.

Baudelaire, Charles. *Œuvres complètes* Ed. Marcel A. Ruff. Paris: Seuil, 1970.

Beaver, Harold. "Doodling America: Poe's 'MS. Found in a Bottle.'" *A Centre of Excellence: Essays Presented to Seymour Betsky.* Ed. Robert Druce. Amsterdam: Rodopi, 1987, 15–27.

Bernard, Suzanne. "Le 'Coup de dés' de Mallarmé replacé dans la perspective historique." *Revue d'histoire littéraire de la France* 51 (April–June 1951): 181–95.

[Bible]. *The Dartmouth Bible* (King James Version). Ed. R. B. Chamberlin and H. Feldman. Boston: Houghton Mifflin, 1961.

Blake, William. *The Complete Poetry and Prose of William Blake.* Ed. David V. Erdman. Commentary by Harold Bloom. New York: Anchor Press, 1988.

Bonaparte, Marie. *The Life and Works of Edgar Allan Poe.* London: Imago, 1949.

Bowie, Malcolm. *Mallarmé and the Art of Being Difficult.* Cambridge: Cambridge University Press, 1978.

——. "The Question of *Un Coup de dés.*" *Baudelaire, Mallarmé, Valéry: New Essays in Honour of Lloyd Austin.* Ed. Malcolm Bowie, Alison Fairlie, and Alison Finch. Cambridge: Cambridge University Press, 1982.

Cambridge Italian Dictionary [*The*]. Ed. Barbara Reynolds. Cambridge: Cambridge University Press, 1962.

Cassell's Italian Dictionary. Compiled by P. Rebora. New York: Macmillan, 1979.

Cassell's Latin Dictionary. By D. P. Simpson. New York: Macmillan, 1968.

Chadwick, Charles. *Mallarmé: sa pensée dans sa poésie.* Paris: Corti, 1962.

Chambers, Ross. "To Read Rimbaud (a) Mimesis and Symbolisation: A Question in Rimbaud Criticism." *Australian Journal of French Studies* 11:1 (January–April 1974): 54–64.

Chevalier, Jean, and Alain Gheerbrant. *Dictionnaire des symboles.* Paris: Laffont, 1969.

Cicero. *On the Commonwealth* [*De Re Publica*]. Trans. G. H. Sabine and S. B. Smith. Indianapolis: Bobbs-Merrill, 1929.

Clark, Kenneth. *The Drawings by Sandro Botticelli for Dante's Divine Comedy.* New York: Harper and Row, 1976.

Cohn, Robert G. *Mallarmé's Masterwork: New Findings.* The Hague: Mouton, 1966.

——. *Mallarmé's* Un Coup de dés, *An exegesis.* New Haven: Yale French Studies, 1949.

Cook, Theodore A. *The Curves of Life.* London, 1914; rpt. New York: Dover, 1979.

Damon, S. Foster. *A Blake Dictionary.* Providence: Brown University Press, 1965.

——. *Blake's Job.* Providence: Brown University Press, 1966.

Dante Alighieri. *The Comedy of Dante Alighieri.* Trans. Dorothy L. Sayers and Barbara Reynolds. 3 vols. Harmondsworth: Penguin, 1974, 1975, 1976.

——. *The Divine Comedy.* Trans. with a commentary by Charles S. Singleton. 3 vols. Princeton: Princeton University Press, 1980.

Davies, Gardner. *Vers une explication rationnelle du* Coup de dés. Paris: Corti, 1953.

Derrida, Jacques. *La Dissémination.* Paris: Seuil, 1972.

Descartes, René. *Œuvres de Descartes.* Ed. Charles Adam and Paul Tannery. Paris: Vrin, 1966. Vol. X.

——. *Œuvres philosophiques.* Ed. Ferdinand Alquié. Paris: Garnier, 1963. Vol. I.

Dictionary of the Bible [*A*]. Ed. James Hastings. New York: Scribner's, 1905. Vol. IV.

Dionysius the Areopagite. *On Divine Names. The Works of Dionysius the Areopagite.* Trans. Rev. John Parker. Merrick, N.Y.: Richwood, 1976.

Eliade, Mircea. *Images et Symboles.* Paris: Gallimard, 1952.

Encyclopaedia Britannica, 11th ed. Cambridge: Cambridge University Press, 1910. Vol. XIV.

Epicurus. *Epicurus: The Extant Remains.* Trans. Cyril Bailey. Oxford: Clarendon Press, 1926.

——. *Letters, Principal Doctrines, and Vatican Sayings.* Trans. Russel M. Geer. Indianapolis: Bobbs-Merrill, 1964.

Felman, Shoshana. "'Tu as bien fait de partir, Arthur Rimbaud': Poésie et modernité." *Littérature* 11 (October 1973): 3–21.

Feuer, Lewis S. "The Dreams of Descartes." *The American Imago* 20:1 (Spring 1963): 3–26.

Figgis, Darrell. *The Paintings of William Blake.* London: Ernest Benn, 1925.

Fowlie, Wallace. *Mallarmé.* Chicago: University of Chicago Press, 1970 (first published 1953).

Freadman, Anne. "To Read Rimbaud (b) A Reading of 'Mystique.'" *Australian Journal of French Studies* 11:1 (January–April 1974): 65–82.

Freccero, John. "Dante's Pilgrim in a Gyre." *PMLA* 76 (1961): 168–81.

——. "Donne's 'Valediction Forbidding Mourning.'" *ELH* 30 (1963): 335–76.

Freeman, Kathleen, trans. and ed. *Ancilla to the Pre-Socratic Philosophers: A Complete Translation of the Fragments in Diels,* [*H.*], Fragmente de Vorsokratiker (ed. with additions by W. Kranz). Oxford: Blackwell, 1948.

Frieden, Ken. "Poe's Narrative Monologues." *Genius and Monologue.* Ithaca, N.Y.: Cornell University Press, 1985. Rpt. in *The Tales of Poe.* Ed. Harold Bloom. New York: Chelsea House, 1987, 135–47.

Friedrich, Hugo. *The Structure of Modern Poetry.* Trans. Joachim Neugroschel. Evanston, Ill.: Northwestern University Press, 1974.

Frye, Northrop. *Fearful Symmetry.* Boston: Beacon Press, 1962.

Galston, Iago. "Descartes and Modern Psychiatric Thought." *Isis* 35 (Spring 1944): 118–28.

Gardner, Martin. *The Ambidextrous Universe.* New York: Basic Books, 1964.

Giusto, Jean-Pierre. *Rimbaud créateur.* Paris: Presses Universitaires de France, 1980.

Gouhier, Henri. *Les Premières pensées de Descartes.* Paris: Vrin, 1958.

Grande Encyclopédie [*La*]. Paris: Société Anonyme de la Grande Encyclopédie, n.d. Vol. XXXI.

Grand Larousse encyclopédique. Paris: Larousse, 1960 and 1964. Vols. III and X.

Grant, John E. "Visions in Vala: A Consideration of Some Pictures in the Manuscript." *Blake's Sublime Allegory.* Ed. Stuart Curran and Joseph A. Wittreich, Jr. Madison: University of Wisconsin Press, 1973, 141–202.

Grimes, Ronald L. "Time and Space in Blake's Major Prophecies." *Blake's Sublime Allegory.* Ed. Stuart Curran and Joseph A. Wittreich, Jr. Madison: University of Wisconsin Press, 1973, 59–81.

Gwynne, Peter with John Carey. "A New Type of Life." *Newsweek,* 24 December 1979, 58.

Halliburton, David. *Edgar Allan Poe: A Phenomenological View.* Princeton: Princeton University Press, 1973.

Harper's Dictionary of Classical Literature and Antiquities. Ed. H. T. Peck. New York: American Book Co., 1923.

Heidegger, Martin. *Being and Time.* Trans. John Macquarrie and Edward Robinson. New York: Harper and Row, 1962.

——. *Identity and Difference.* Trans. Joan Stambaugh. New York: Harper and Row, 1969.

Hobhouse, L. T. *Morals in Evolution: A Study in Comparative Ethics.* London: Chapman and Hall, 1951.

Homer. *The Iliad of Homer.* Trans. Richmond Lattimore. Chicago: University of Chicago Press, 1974.

——. *The Odyssey.* Trans. Robert Fitzgerald. Garden City, N.Y.: Doubleday, 1963.

Johansen, Svend. "Le Problème d'*Un Coup de dés.*" *Orbis Litterarum* 3:1 (1945): 282–313.

Judson, Horace F. *The Search for Solutions.* New York: Holt, Rinehart and Winston, 1980.

Kennington, Richard. "Descartes' 'Olympica.'" *Social Research* 28 (July 1961): 171–204.

Kirk, G. S., and J. E. Raven. *The Presocratic Philosophers: A Critical History with a Selection of Texts.* Cambridge: Cambridge University Press, 1977.

Kittang, Atle. *Discours et jeu: essai d'analyse des textes d'Arthur Rimbaud.* Bergen, Norway: Universitetsførlaget, 1975.

La Charité, Virginia A. *The Dynamics of Space: Mallarmé's* Un Coup de dés jamais n'abolira le hasard. Lexington, Ky.: French Forum, 1987.

Lapeyre, Paule. *Le Vertige de Rimbaud: clé d'une perception poétique.* Neuchâtel: La Baconnière, 1981.

Larousse, Pierre. *Grand Dictionnaire universel.* Paris: Administration du Grand Dictionnaire universel, n.d. Vol. XV.

Leroy, Maxime. *Descartes, le philosophe au masque.* Paris: Rieder, 1929. Vol. I.

Lowes, John Livingston. *The Road to Xanadu, A Study in the Ways of the Imagination.* Boston: Houghton Mifflin, 1927.

[Lucretius]. Titi Lucreti Cari. *De Rerum Natura* (Libri Sex). Ed. and trans. Cyril Bailey. 3 vols. Oxford: Clarendon Press, 1949.

Lucretius. *The Way Things Are [De Rerum Natura]*. Trans. Rolfe Humphries. Bloomington: Indiana University Press, 1969.
Luther, Martin. *The Bondage of the Will*, in *Martin Luther: Selections from His Writings*. Ed. John Dillenberger. Garden City, N.Y.: Anchor Books, 1961.
Mackenzie, Donald A. *The Migration of Symbols*. New York: Knopf, 1926.
[Mallarmé, Stéphane]. *Mallarmé*. Ed. and trans. Anthony Hartley. Harmondsworth: Penguin, 1970.
Mallarmé, Stéphane. *Œuvres complètes*. Ed. Henri Mondor and G. Jean-Aubry. Paris: Gallimard, 1945.
——. *The Poems*. Trans. Keith Bosley. Harmondsworth: Penguin, 1977.
——. *Selected Poetry and Prose*. Ed. Mary Ann Caws. New York: New Directions, 1982.
——. *Un Coup de dés jamais n'abolira le hasard*. Ed. Mitsou Ronat. Paris: Change errant/d'atelier, 1980.
McGraw-Hill Encyclopedia of Science and Technology. New York: McGraw-Hill, 1977. Vol. XIV.
Mellor, Anne K. *Blake's Human Form Divine*. Berkeley: University of California Press, 1974.
Meltzer, Françoise. "Descartes' Dreams and Freud's Failure, or The Politics of Originality." *The Trial(s) of Psychoanalysis*. Ed. Françoise Meltzer. Chicago: Chicago University Press, 1988, 81–102.
Milton, John. *Complete Poems and Major Prose*. Ed. Merritt Y. Hughes. Indianapolis: Odyssey Press, 1978.
Mitchell, W.J.T. *Blake's Composite Art*. Princeton: Princeton University Press, 1978.
Molière. *Les Femmes sçavantes*. *Œuvres complètes de Molière*. Ed. René Bray. Paris: Les Belles Lettres, 1952.
——. *The Learned Ladies*. Trans. Renée Waldinger. Woodbury, N.Y.: Barron's, 1957.
Mondor, Henri. *Vie de Mallarmé*. Paris: Gallimard, 1941–42.
Mooney, Stephen L. "Poe's Gothic Waste Land." *Sewanee Review*, January–March 1962. Rpt. in *The Recognition of Edgar Allan Poe: Selected Criticism Since 1829*. Ed. Eric W. Carlson. Ann Arbor: University of Michigan Press, 1970, 278–97.
Morton III, W. Brown, and Dean Conger. "Indonesia Rescues Ancient Borobudur." *National Geographic Magazine*, 163:1, January 1983, 126–42.
Nurmi, Martin K. *William Blake*. London: Hutchinson University Library, 1975.
Oxford English Dictionary [The Compact Edition of the]. Complete Text Reproduced Micrographically. 2 vols. Oxford: Oxford University Press, 1979.
Peithman, Stephen. *The Annotated Tales of Edgar Allan Poe*. Garden City, N.Y.: Doubleday, 1981.
Plato. *Plato's Cosmology: The Timaeus of Plato*. Trans. with a running commentary by Francis M. Cornford. Indianapolis: Bobbs-Merrill, 1975.
——. *The Republic*. Trans. Paul Shorey. 2 vols. Cambridge: Harvard University Press, 1956.
——. *The Republic of Plato*. Trans. Francis M. Cornford. London: Oxford University Press, 1945.
——. *Timaeus*. Trans. B. Jowett. Ed. G. R. Morrow. Indianapolis: Bobbs-Merrill, 1949.
Plessen, Jacques. "*Marine* de Rimbaud: une analyse." *Neophilologus* 60:1 (January 1971): 16–32.
Poe, Edgar A. *The Complete Works of Edgar Allan Poe*. Ed. James A. Harrison. New York: AMS Press, 1965. Vols. II and XVI.
Pontoppidan, Erich. *The Natural History of Norway*. London, 1755. Vol. I.
Poulet, Georges. *Études sur le temps humain*. Edinburgh: University Press, 1949.
Purce, Jill. *The Mystic Spiral, Journey of the Soul*. New York: Thames and Hudson, 1980.
Raine, Kathleen. *Blake and Tradition*. Princeton: Princeton University Press, 1968.

Ramsey, Warren. "A View of Mallarmé's Poetics." *The Romanic Review* 46:3 (October 1955): 178–91.

Rimbaud, Arthur. *Complete Works, Selected Letters*. Trans. Wallace Fowlie. Chicago: University of Chicago Press, 1966.

——. *Illuminations and Other Prose Poems*. Trans. Louise Varèse. New York: New Directions, 1957.

——. *Œuvres*. Ed. Suzanne Bernard. Paris: Garnier, 1960.

Robert, Paul. *Dictionnaire alphabétique et analogique de la langue française*. Paris: Le Robert, 1986.

Roe, Albert S. *Blake's Illustrations to the Divine Comedy*. Princeton: Princeton University Press, 1953.

Roulet, Claude. *Éléments de poétique mallarméenne d'après le poème* Un Coup de dés jamais n'abolira le hasard. Neuchâtel: Editions du Griffon, 1947.

Sartre, Jean-Paul. *L'Être et le néant*. Paris: Gallimard, 1943.

Scherer, Jacques. *L'Expression littéraire dans l'œuvre de Mallarmé*. Paris: Nizet, 1947.

Schönberger, Stephen. "A Dream of Descartes: Reflections on the Unconscious Determinants of the Sciences." *The International Journal of Psycho-Analysis* 20 (January 1939): 43–57.

Sebba, Gregor. *The Dream of Descartes*. Ed. Richard A. Watson. Carbondale, Ill.: Southern Illinois University Press, 1987.

Serres, Michel. *La Naissance de la physique dans le texte de Lucrèce: fleuves et turbulences*. Paris: Éditions de minuit, 1977.

Shipman, Harry L. *Black Holes, Quasars, and the Universe*. Boston: Houghton Mifflin, 1976.

Shulman, Robert. "Poe and the Powers of the Mind." *ELH*, 37 (1970). Cited by Gerard M. Sweeney. "Beauty and Truth: Poe's 'A Descent into the Maelström.'" *Poe Studies* 6 (June 1973): 22–25.

Simon, Gérard. "Descartes, le rêve et la philosophie au XVIIe siècle." *Revue des sciences humaines* 82:211 (July–September 1988): 133–51.

Stauffer, Donald B. "The Two Styles of Poe's 'MS. Found in a Bottle.'" *Style* I:2 (Spring 1967): 107–20.

Stiskin, Nahum. *The Looking Glass God . . . Shinto, Yin-Yang, and a Cosomology for Today*. Tokyo: Autumn Press, 1972.

Sweeney, Gerard M. "Beauty and Truth: Poe's 'A Descent into the Maelström.'" *Poe Studies* 6 (June 1973): 22–25.

Thewlis, J., ed. *Encyclopædic Dictionary of Physics*. Oxford: Pergamon Press, 1962. Vol. VII.

Thompson, D'Arcy Wentworth. *On Growth and Form*, 2nd ed. 1942; rpt. London: Cambridge University Press, 1979. Vol. II.

Todorov, Tzvetan. "Une Complication de texte: les 'Illuminations.'" *Poétique* 34 (April 1978): 241–53.

Van Nostrand's Scientific Encyclopedia. Princeton: Van Nostrand, 1958.

von Franz, Marie-Louise. "The Dream of Descartes." *Timeless Documents of the Soul*. Evanston, Ill.: Northwestern University Press, 1968, 55–147.

Vrooman, Jack. *René Descartes, A Biography*. New York: Putnam's, 1970.

Walzer, Pierre-Olivier. *Essai sur Mallarmé*. Paris: Seghers, 1963.

Watson, James D. *The Double Helix*. New York: Atheneum, 1969.

Wilbur, Richard. "The House of Poe" (Library of Congress Anniversary Lecture, 4 May 1959). Rpt. in *The Recognition of Poe: Selected Criticism Since 1829*. Ed. Eric W. Carlson. Ann Arbor: University of Michigan Press, 1970, 255–77.

Wilkie, Brian, and Mary Lynn Johnson. *Blake's Four Zoas: The Design of a Dream.* Cambridge: Harvard University Press, 1978.

Williams, Thomas. *Mallarmé and the Language of Mysticism.* Athens: University of Georgia Press, 1970.

Yeats, W. B. *A Vision.* New York: Collier, 1969.

Index of Sigla

AR Abell, George. *Realm of the Universe.*
AV Aiton, E. J. *The Vortex Theory of Planetary Motions.*
B *The Dartmouth Bible* (King James Version).
BFZ Blake, William. *The Four Zoas,* in *The Complete Poetry and Prose of William Blake.*
BG Bailey, Cyril. *The Greek Atomists and Epicurus: A Study.*
BP Bachelard, Gaston. *La Poétique de l'espace.*
BV Baillet, Adrien. *La Vie de Monsieur Des-cartes.*
CC Cook, Theodore A. *The Curves of Life.*
CI *Cassell's Italian Dictionary.*
CM Cohn, Robert G. *Mallarmé's* Un Coup de Dés, *an exegesis.*
DB Damon, S. Foster. *A Blake Dictionary.*
DJ Damon, S. Foster. *Blake's Job.*
DO Descartes, René. *Œuvres de Descartes.* Ed. Charles Adam and Paul Tannery.
DSA Dante Alighieri. *The Comedy of Dante Alighieri.* Trans. Dorothy L. Sayers and Barbara Reynolds.
DSI ——. *The Divine Comedy.* Trans. Charles S. Singleton.
EI Eliade, Mircea. *Images et Symboles.*
FB Figgis, Darrell. *The Paintings of William Blake.*
FD Freeman, Kathleen, trans. and ed. *Ancilla to the Pre-Socratic Philosophers: A Complete Translation of the Fragments in Diels,* [*H.*], *Fragmente de Vorsokratiker.*
FF Frye, Northrop. *Fearful Symmetry.*
FG Freccero, John. "Dante's Pilgrim in a Gyre."
GA Gardner, Martin. *The Ambidextrous Universe.*
GD Gouhier, Henri. *Les Premières pensées de Descartes.*
GE *La Grande Encyclopédie.*
GL *Grand Larousse encyclopédique.*
GR Giusto, Jean-Pierre. *Rimbaud créateur.*
GT Grimes, Ronald L. "Time and Space in Blake's Major Prophecies."
HO Homer. *The Odyssey.* Trans. Robert Fitzgerald.
HP Halliburton, David. *Edgar Allan Poe: A Phenomenological View.*
JS Judson, Horace F. *The Search for Solutions.*
KD Kennington, Richard. "Descartes' 'Olympica.'"
KR Kirk, G. S., and J. E. Raven. *The Presocratic Philosophers: A Critical History with a Selection of Texts.*

LG Larousse, Pierre. *Grand Dictionnaire universel.*

LR Lucretius. *De Rerum Natura.* Ed. and trans. Cyril Bailey.

M Mallarmé, Stéphane. *Œuvres complètes.* Ed. Henri Mondor and G. Jean-Aubry.

MB Mitchell, W.J.T. *Blake's Composite Art.*

MH *McGraw-Hill Encyclopedia of Science and Technology.*

MM Mackenzie, Donald A. *The Migration of Symbols.*

OED *Oxford English Dictionary* [*The Compact Edition of the*].

PE Poulet, Georges. *Études sur le temps humain.*

PM Purce, Jill. *The Mystic Spiral, Journey of the Soul.*

PRC Plato. *The Republic of Plato.* Trans. Francis M. Cornford.

PRS ——. *The Republic.* Trans. Paul Shorey.

PTC ——. *Plato's Cosmology: The Timaeus of Plato.* Trans. with a running commentary by Francis M. Cornford.

PTJ ——. *Timaeus.* Trans. B. Jowett. Ed. G. R. Morrow.

R Rimbaud, Arthur. *Œuvres.* Ed. Suzanne Bernard.

RB Roe, Albert S. *Blake's Illustrations to the Divine Comedy.*

RT Raine, Kathleen. *Blake and Tradition.*

SB Shipman, Harry L. *Black Holes, Quasars, and the Universe.*

SL Stiskin, Nahum. *The Looking Glass God . . . Shinto, Yin-Yang, and a Cosomology for Today.*

SN Serres, Michel. *La Naissance de la physique dans le texte de Lucrèce: fleuves et turbulences.*

TC Todorov, Tzvetan. "Une Complication de texte: les 'Illuminations.'"

TE Thewlis, J., ed. *Encyclopædic Dictionary of Physics.*

TG Thompson, D'Arcy Wentworth. *On Growth and Form.*

VN *Van Nostrand's Scientific Encyclopedia.*

Index

Anaxagoras, 22–23, 27, 38, 47, 144, 173 nn. 11, 13, 18
compared to atomists, 24–25
vortex in, 22–23
Anaximander, 17
Anaximenes, 16
Apollinaire, Guillaume, 187 n. 8
Archimedes, 44, 110, 155, 175 n. 12
Aristotle, 16, 17, 20, 25, 27, 38, 53, 173 nn. 11, 20, 175 n. 12
atomists (*see also* Leucippus; Democritus), 34, 47, 144, 173 n. 19
vortex in, 24–25, 37–39
Augustine, Saint, 61, 75
whirlpool (whirl) in, 54–55, 56

Bachelard, Gaston, 6–7
Baudelaire, Charles, 115, 141, 183 n. 1
Works:
"Le Gouffre," 102, 133
"Le Voyage," 141
Beckett, Samuel, 138
Bible, 47, 85, 92, 94
New Testament, 10, 88
Old Testament, 9–11, 47, 67, 88, 89, 144, 146, 181 n. 23
whirlwind in, 9–11, 89, 172 n. 19
black hole, 104, 106, 161–62
Blake, William, 85–97, 101, 112, 127, 145, 146, 147, 189 n. 26
comparison of, with Dante, 108
spiral in, 87, 88, 90, 95, 181–82 n. 2
spiral helix in, 88, 89, 90, 92, 95
vortex in, 85–97, 124, 181–82 n. 2
whirlwind in, 88
whorl in, 94–95
Works:
"The Epitome of James Hervey's Meditations among the Tombs," 92
Europe, 87
"Ezekiel's Vision," 92
The Four Zoas, 86, 172 n. 20, 182 nn. 10, 20
illustrations to the Book of Job, 88–89, 90, 95
illustrations to Dante's *Comedy,* 89, 93–94, 95
illustrations to *Paradise Lost,* 92
"Jacob's Ladder," 92
Milton, 90–92
Newton, 87
"The Plow of Jehovah" (*Jerusalem*), 90, 93, 95
Botticelli, Sandro, 51, 171 n. 8
"Boulak Papyrus," 47, 144
whirlpool in, 8–9
Breton, André, 115

Cicero, 174 nn. 5, 6, 177 n. 20
"Scipio's Dream," 174 nn. 4, 8, 11, 181 n. 20
"circumstance," 48
analysis of, by Serres, 45

"circumstance" (*continued*)
in Dante, 56
in Mallarmé, 135–36
in Plato, 45
in Poe, 112
Coleridge, Samuel Taylor
"The Rime of the Ancient Mariner," 105, 107, 183 n. 7
Coriolis, Gaspard Gustave de, 168, 169, 191 n. 32

Dante Alighieri, 51–69, 85, 89, 93, 94, 95, 96, 97, 101, 140, 145, 146, 157, 174 n. 7, 176 n. 11, 177 nn. 14, 20, 183 n. 7, 189 n. 26
"circumstance" in, 56
comparison of, with Blake, 108
comparison of, with Poe, 103–4, 105–6
enantiomorphism in, 59
gyre in, 51–52, 62, 65–66
helix in, 57–59
spiral in, 66, 176 n. 12
spiral helix in, 57, 59, 61, 66, 176 n. 11, 177 n. 16
vertigo in, 55
vortex in, 52–59 (*Inferno*), 59–62 (*Purgatorio*), 62–69 (*Paradiso*)
whirlpool in, 58
whirlwind in, 55–57, 67
Works:
Commedia, 48, 52, 54, 57, 59, 68, 69, 94
Inferno, 53, 54, 57, 93
Paradiso, 65, 94
Purgatorio, 54
Darwin, Charles, 163
Democritus (*see also* atomists), 24–25, 39, 46, 112, 173 nn. 19, 20
Descartes, René, 71–84, 85, 95, 96, 97, 101, 112, 145–46, 150, 178 n. 8, 178–79 n. 9, 179 n. 13, 179–80 n. 14, 180 n. 20, 181 nn. 22, 24
mise en abyme in, 73
vertigo in, 82–83, 181 n. 25
vortex in, 71–84, 87, 91, 177 n. 1
vortex of, in Molière, 178 n. 1
whirlwind in, 75–84, 181 n. 23
Works:
Cogitationes Privatae, 180 n. 15
Discours de la méthode, 84
Méditations, 84, 180 n. 18
Le Monde, ou traité de La Lumière, 177 n. 1
Olympica, 72, 82, 179 n. 13, 180 n. 15, 181 nn. 20, 24, 27
Principia philosophiae, 71, 87
Dionysius the Areopagite, 182 nn. 23, 24

Einstein, Albert, 161, 162
Eliade, Mircea, 6–7
Elijah, 10, 11, 82, 172 n. 19
Empedocles, 17, 19–22, 23, 32, 45, 47, 144, 173 nn. 11, 13
enantiomorphism in, 21
vortex in, 19–21
enantiomorphism, 162–69, 190 n. 22
in Dante, 59
in Empedocles, 21
helical, 165–66
spiralic, 165
spiro-helical, 167–68
vortical, 168–69
Epicurus, 37–39, 42, 43, 44, 46, 47, 48, 144, 173 n. 19
swerve in, 38, 46, 48
vortex in, 37–39, 175 n. 7
Ezekiel, 82, 94, 172 n. 19
in Bible, 10–11
in Blake, 92–93

Freud, Sigmund, 72, 181 n. 26

Galileo, 72, 84, 177 n. 1
Gide, André, 138, 187 n. 8
gyre, 143
in Dante, 51–52, 62, 65–66
defined, 157
W. B. Yeats's analysis of, in Empedocles, 20–21

Heidegger, Martin, 83, 154, 181 n. 25
helix, 153–56, 163–64, 190 nn. 7, 10, 24, 191 n. 27
in Dante, 57–59
defined, 154
enantiomorphic, 165–66
in Mallarmé, 189 n. 26
Heraclitus, 16–18, 21–22, 23, 47, 144, 173 n. 13, 176 n. 13, 185 n. 8
vortex in, 16–17

Iliad (Homer)
vortical symbolism in, 11

Job, 10, 172 n. 19
in Blake, 88, 90, 95

Keats, John
"Ode to a Nightingale," 75–76
Kepler, Johann, 72

Leucippus (*see also* atomists), 24–25, 173 n. 19
Lowes, John Livingston, ix
Lucretius, 38–46, 47, 48, 123, 137, 139, 144–45, 147, 173 n. 19, 176 n. 18, 185 n. 8
swerve in, 43–44, 45–46, 48
swerve of, in Mallarmé, 137, 139
vertigo in, 42
vortex in, 38–46
waterspout in, 43, 176 n. 13
Works:
De Rerum Natura, 42, 51, 136, 173 n. 7, 175 n. 10, 189 n. 23
Luther, Martin
The Bondage of the Will, 81

maelstrom. *See* vortex; whirlpool
Mallarmé, Stéphane, 101, 129–39, 146, 147, 186 n. 3, 187 n. 5, 187–88 n. 8, 188 n. 19
"circumstance" in, 135–36
helix in, 189 n. 26
Lucretian swerve in, 137, 139
spiral in, 139, 189 n. 26
spiral helix in, 189 n. 26
vertigo in, 139, 189 n. 21
vortex in, 129–39, 141, 186 n. 1, 188 n. 9, 189 nn. 21, 26
whirlpool in, 136, 137, 189 n. 21
Works:
Un Coup de dés jamais n'abolira le hasard, 129–39, 143, 147, 186 n. 1, 187 n. 5, 187–88 n. 8, 188 nn. 13, 19, 189 nn. 23, 26
"Billet à Whistler," 186 n. 1
"Brise Marine," 186 n. 1
Igitur, 135, 139, 186 n. 1, 189 n. 26
"A la nue accablante tu," 186 n. 1
"Salut," 186 n. 1
"Le Tombeau d'Edgar Poe," 141
Milton, John, 85, 97
in Blake, 91, 92
Paradise Lost, 87, 182 n. 10
mise en abyme
in Descartes, 73
in Poe, 107
Molière
Descartes's vortex in, 178 n. 1
Les Femmes sçavantes, 178 n. 1

Neoplatonism, 95
Newton, Isaac, 87, 90–91
Nietzsche, Friedrich, 83

Odyssey (Homer), 47, 56, 144
Skylla and Kharybdis, 12–14, 183–84 n. 13
vortical symbolism in, 11–14

Parmenides, 34
Pascal, Blaise, 102
Pasteur, Louis, 190 n. 22
Petrarch, 75
phenomenology, ix–x, 23, 111–12
Plato, 27–36, 44–45, 47, 172 n. 2, 173 n. 13, 174 n. 11, 176 n. 18, 177 n. 20
"circumstance" in, 45
"Great Whorl" of, 27–36, 53, 62–63, 144, 174 nn. 4, 6, 177 n. 24
"Myth of Er," 27–36
spiral in, 32
vortex in, 27–36
Works:
Phaedo, 27
Republic, 27–36
Timaeus, 11, 28–36, 172 n. 21
Poe, Edgar Allan, 101–13, 115, 116, 133, 141, 146, 147, 183 nn. 1, 11, 13
"circumstance" in, 112
comparison of, with Dante, 103–4, 105–6
mise en abyme in, 107
spiral in, 102, 103, 111, 112
spiral helix in, 112
vertigo in, 109, 183 n. 8
vortex in, 102–13, 140, 184 n. 16
whirlpool in, 102, 103–11, 113, 124, 183 n. 8
whirlwind in, 102
Works:
"A Descent into the Maelström," 102, 107–10, 111, 112, 146
Eureka, 112
"The Fall of the House of Usher," 102
"King Pest," 102
"Metzengerstein," 102
"MS. Found in a Bottle," 102–7, 108, 109, 183 nn. 6, 10

Poulet, Georges, 74, 75, 79, 81–82, 180 nn. 19, 20
pyramid, 59
Pythagoras, 75

Rimbaud, Arthur, 101, 115–28, 129, 130, 146–47, 184 n. 7, 185 n. 16, 186 nn. 3, 22
seashell in, 125, 126, 185 n. 19
spiral helix in, 124, 126
vertigo in, 120–21, 122, 123, 127, 185 nn. 15, 19
vortex in, 116–28, 140–41, 185 n. 8
waterspout in, 116–17, 120–21, 122, 123
whirlpool in, 117, 119, 123
whirlwind in, 117
Works:
"Le Bal des pendus," 185 n. 9
"Le Bateau ivre," 116–17, 122, 127, 185 n. 8
"Les Etrennes des orphelins," 117
Illuminations, 115, 127, 140, 146, 185 n. 13
"Marine," 117–19, 122, 123, 185 n. 10
"Matinée d'ivresse," 185 n. 19
"Mouvement," 117, 120–23
"Mystique," 117, 124–27
"Qu'est-ce pour nous, mon coeur," 117, 127

Sartre, Jean-Paul, 83, 138, 181 n. 25
seashell (*coquille*), 8, 95, 144, 150, 151, 152, 163, 167, 171 nn. 5, 8, 10, 171–72 n. 11
in Lucretius, 41
in Rimbaud, 125, 126, 185 n. 19
symbolism of, 6–7
Serres, Michel, ix, 15, 56, 112, 123, 135, 137, 175 n. 6, 176 n. 18
"circumstance" analyzed by, 45
on turbulence in Lucretius, 38–46, 175 n. 12
Shakespeare, William, 139
spiral, 149–53, 154, 155, 157, 160, 163, 164, 166, 171 n. 5, 175 n. 12, 189 nn. 3, 5, 189–90 n. 6, 190 nn. 9, 10, 24, 190–91 n. 26
Archimedean, 5, 149
in Blake, 87, 88, 90, 95, 181–82 n. 2
in Dante, 66, 176 n. 12
defined, 149–52
enantiomorphic, 165
equable, 149
equiangular, 5, 150
logarithmic, 5, 150, 152, 153, 158, 171 n. 10, 190 nn. 10, 26, 30
in Mallarmé, 139, 189 n. 26
in Plato, 32
in Poe, 102, 103, 111, 112
symbolism of, 4–5, 96, 171 n. 3
spiral helix, 6, 11, 96, 156–57, 164, 166, 167, 168, 171 nn. 5, 9, 11, 190 nn. 10, 24, 190–91 n. 26
in Blake, 88, 89, 90, 92, 95
in Dante, 57, 59, 61, 66, 176 n. 11, 177 n. 16
defined, 156
enantiomorphic, 167–68
in Mallarmé, 189 n. 26
in Poe, 112
in Rimbaud, 124, 126
swerve (*clinamen*)
in Epicurus, 38, 46, 48
in Lucretius, 43–44, 45–46, 48
of Lucretius in Mallarmé, 137, 139

Tao symbol (yin-yang), 3–5, 166, 191 n. 30
Thales, 16
Todorov, Tzvetan, 184 n. 7, 186 n. 3
on Rimbaud, 115–16
turbulence, 8, 118, 140, 146–47, 157, 167–69
ancient symbols of, 3
description of, ix
in Heraclitus, 17
in Poe, 102–3
symbolic (aesthetic), ix, 4, 47–48, 143, 144, 149, 169

Verlaine, Paul, 124, 188 n. 19
vertigo, 96–97
in Dante, 55
in Descartes, 82–83, 181 n. 25
in Lucretius, 42
in Mallarmé, 139, 189 n. 21
in Poe, 109, 183 n. 8
in Rimbaud, 120–21, 122, 123, 127, 185 nn. 15, 19
vortex (*tourbillon*), ix, 7–8, 157–62, 167–69, 173 n. 10, 190 n. 13
in Anaxagoras, 22–23
in ancient texts, 47–48
in atomists, 24–25
in Bible, 82
in Blake, 85–97, 124, 181–82 n. 2
bound, 159, 167
in Dante, 52–59 (*Inferno*), 59–62 (*Purgatorio*), 62–69 (*Paradiso*)
defined, 158–59

vortex (*continued*)
in Descartes, 71–84, 87, 91, 177 n. 1
of Descartes in Molière, 178 n. 1
in Empedocles, 19–21
enantiomorphic, 168–69
in Epicurus, 37–39, 175 n. 7
forced, 63, 65, 158, 161
free, 158, 160
free circular, 27, 33, 52, 60, 63, 65, 67, 158
free spiral, 158
in Heraclitus, 16–17
in Homer, 11–14
in Lucretius, 38–46
in Mallarmé, 129–39, 141, 186 n. 1, 188 n. 9, 189 nn. 21, 26
in Plato, 27–36
in Poe, 102–13, 140, 184 n. 16
in Rimbaud, 116–28, 140–41, 185 n. 8
stationary, 159
symbolism of, 143–48
Vorticism, 143

waterspout (*prester*), 158, 160, 169, 173 n. 7, 185 n. 8
fiery, in Heraclitus, 16–17, 47
in Lucretius, 43, 176 n. 13
in Rimbaud, 116–17, 120–21, 122, 123
whirl, 145, 148, 175 n. 11, 181 n. 25
in Anaxagoras, 22–23
in atomists, 24–25
in Augustine, 55
in Blake, 95
in Empedocles, 19–20
in Epicurus, 37–39, 175 n. 7
in Homer, 11–14
in Mallarmé, 186 n. 1
in Rimbaud, 117, 126
symbolism of, 47, 140
whirlpool (maelstrom), 144, 158, 171 n. 8
in Augustine, 54, 56
in "Boulak Papyrus," 8–9
in Dante, 58
in Mallarmé, 136, 137, 189 n. 21
in *Odyssey*, 12–14
in Poe, 102, 103–11, 113, 124, 183 n. 8
in Rimbaud, 117, 119, 123
symbolism of, 8–9
whirlwind, 144, 146, 158, 160, 172 n. 12
in Bible, 9–11, 89, 172 n. 19
in Blake, 88
in Dante, 55–57, 67
in Descartes, 75–84, 181 n. 23
in Poe, 102
in Rimbaud, 117
whorl, 150, 152, 167
in Blake, 94–95
Plato's "Great Whorl," 27–36, 53, 62–63, 174 nn. 4, 6, 177 n. 24
of Tao symbol, 3–5

Yeats, W. B., 143
on Empedocles' antithetical gyres, 20–21
Works:
"Sailing to Byzantium," 157
A Vision, 173 n. 16

ziggurat, 59, 176–77 n. 14